Tassilo Keilmann

Strongly correlated quantum physics with cold atoms

Tassilo Keilmann

Strongly correlated quantum physics with cold atoms

Fractional statistics in optical lattices,
supersolid phases, spin-charge separation
and non-abelian anyons engineered by
ultracold atoms

Südwestdeutscher Verlag für Hochschulschriften

Impressum/Imprint (nur für Deutschland/ only for Germany)
Bibliografische Information der Deutschen Nationalbibliothek: Die Deutsche Nationalbibliothek verzeichnet diese Publikation in der Deutschen Nationalbibliografie; detaillierte bibliografische Daten sind im Internet über http://dnb.d-nb.de abrufbar.

Alle in diesem Buch genannten Marken und Produktnamen unterliegen warenzeichen-, marken- oder patentrechtlichem Schutz bzw. sind Warenzeichen oder eingetragene Warenzeichen der jeweiligen Inhaber. Die Wiedergabe von Marken, Produktnamen, Gebrauchsnamen, Handelsnamen, Warenbezeichnungen u.s.w. in diesem Werk berechtigt auch ohne besondere Kennzeichnung nicht zu der Annahme, dass solche Namen im Sinne der Warenzeichen- und Markenschutzgesetzgebung als frei zu betrachten wären und daher von jedermann benutzt werden dürften.

Verlag: Südwestdeutscher Verlag für Hochschulschriften Aktiengesellschaft & Co. KG
Dudweiler Landstr. 99, 66123 Saarbrücken, Deutschland
Telefon +49 681 37 20 271-1, Telefax +49 681 37 20 271-0
Email: info@svh-verlag.de
Zugl.: München, LMU, Diss., 2009

Herstellung in Deutschland:
Schaltungsdienst Lange o.H.G., Berlin
Books on Demand GmbH, Norderstedt
Reha GmbH, Saarbrücken
Amazon Distribution GmbH, Leipzig
ISBN: 978-3-8381-1311-1

Imprint (only for USA, GB)
Bibliographic information published by the Deutsche Nationalbibliothek: The Deutsche Nationalbibliothek lists this publication in the Deutsche Nationalbibliografie; detailed bibliographic data are available in the Internet at http://dnb.d-nb.de.

Any brand names and product names mentioned in this book are subject to trademark, brand or patent protection and are trademarks or registered trademarks of their respective holders. The use of brand names, product names, common names, trade names, product descriptions etc. even without a particular marking in this works is in no way to be construed to mean that such names may be regarded as unrestricted in respect of trademark and brand protection legislation and could thus be used by anyone.

Publisher: Südwestdeutscher Verlag für Hochschulschriften Aktiengesellschaft & Co. KG
Dudweiler Landstr. 99, 66123 Saarbrücken, Germany
Phone +49 681 37 20 271-1, Fax +49 681 37 20 271-0
Email: info@svh-verlag.de

Printed in the U.S.A.
Printed in the U.K. by (see last page)
ISBN: 978-3-8381-1311-1

Copyright © 2010 by the author and Südwestdeutscher Verlag für Hochschulschriften Aktiengesellschaft & Co. KG and licensors
All rights reserved. Saarbrücken 2010

Abstract

This thesis is devoted to exploit strong correlations among ultracold atoms in order to create novel, exotic quantum states. In the first two chapters, we devise dynamical out-of-equilibrium preparation schemes which lead to intriguing final states.

First of all, we propose to create the elusive **supersolid state** via a quantum quench protocol. Supersolids – quantum hybrids exhibiting both superflow and solidity – have been envisioned long ago, but have not been demonstrated in experiment so far. Our proposal to create a supersolid state is perfectly accessible with current technology and may clear the way to the experimental observation of supersolidity.

Another out-of-equilibrium preparation scheme is discussed in the second chapter, giving rise to novel **Cooper pairs of bosons**. Ordinarily, Cooper pairs are composed of fermions – not so in our setup! We show that a Mott state of local bosonic Bell pairs can evolve into a Cooper-paired state of bosons, where the size of the pairs becomes macroscopic. This state can be prepared via a quick, diabatic ramp from the Mott into the superfluid regime.

Furthermore, we propose to use bosons featuring conditional-hopping amplitudes in order to create **Abelian anyons** in one-dimensional optical lattices. We derive an exact mapping between anyons and bosons via a "fractional" Jordan-Wigner transformation. We suggest to employ a laser-assisted tunneling scheme to establish the many-particle state of "conditional-hopping bosons", thus realizing a gas of Abelian anyons. The fractional statistics phase can be directly tuned by the lasers.

The realization of **non-Abelian anyons** would be especially delightful, due to their significance in topological quantum computation schemes. We propose to employ strongly correlated bosons in one-dimensional optical lattices to create the Pfaffian state – which is known to host non-Abelian anyons as elementary excitations. In this setup, three-body correlations are required to dominate the system, which we model by on-site interactions of atoms with diatomic molecules.

Finally, we use strong correlations in one-dimensional systems to create the effect of **spin-charge separation**, as formulated theoretically first in 1968. For a model of two-component bosons we compute the effective mass of a spin-flip excitation via Bethe Ansatz. In the strongly correlated regime, we show that the effective mass reaches the total mass of all particles in the system. The spin wave thus travels much more slowly than the density wave, giving rise to a strong spin-charge separation.

Zusammenfassung

Diese Arbeit widmet sich der Erzeugung neuartiger, exotischer Quantenzustände durch stark korrelierte ultrakalte Atome. Als Erstes zeigen wir, wie der sogenannte **suprasolide Zustand** in einem optischen Gitter erzeugt werden kann, indem äußere System-Parameter plötzlich verändert werden. Suprasolidität bezeichnet einen neuartigen Materie-Zustand, in dem sich die Atome gleichzeitig sowohl in der festen als auch in der suprafluiden Phase befinden. Eine solche suprasolide Phase wurde bislang experimentell nicht nachgewiesen. Unser Vorschlag, einen suprasoliden Zustand dynamisch zu erzeugen, ist mit gegenwärtiger experimenteller Technik kompatibel und könnte den Weg zum ersten Nachweis der Suprasolidität bereiten.

Im zweiten Kapitel beschreiben wir eine weitere dynamische Methode, um neuartige **Cooper-Paare aus Bosonen** in optischen Gittern zu generieren. Das Konzept der Cooper-Paare, die normalerweise aus antikorrelierten Fermionen bestehen, wird somit auf Bosonen übertragen. Ausgehend von einem Mott-Zustand aus lokalen Bell-Paaren zeigen wir, wie daraus bosonische Cooper-Paare entstehen können. Dazu ist lediglich ein schneller, diabatischer Übergang vom Mott-Regime in das suprafluide Regime nötig.

Desweiteren befassen wir uns mit der Herstellung **Abelscher Anyonen** in optischen Gittern. Wir beweisen, dass Anyonen in einer räumlichen Dimension exakt auf Bosonen abgebildet werden können, wenn deren Tunnelrate von der Besetzung durch andere Bosonen abhängt. Wir beschreiben eine Methode, mit mehreren Raman-Übergängen ein solches System aus "conditional-hopping" Bosonen zu implementieren, was letztlich der Realisierung von Anyonen gleichkäme. Die Austausch-Phase, die die fraktionale Statistik der Anyonen bestimmt, kann durch die Raman-Laser einfach eingestellt werden.

In einem weiteren Kapitel befassen wir uns mit **nicht-Abelschen Anyonen**, deren experimenteller Nachweis besonders reizvoll wäre. Wir zeigen, wie stark korrelierte Bosonen in eindimensionalen optischen Gittern präpariert werden müssen, um den sogenannten Pfaffschen Grundzustand anzunehmen. Elementare Anregungen dieses Zustands können mit nicht-Abelschen Anyonen identifiziert werden. Um den Pfaffschen Zustand zu erzeugen, müssen Dreikörper-Wechselwirkungen – die sonst nur selten in der Natur vorkommen – alle anderen Parameter des Systems dominieren. Wir zeigen wie solche Dreikörper-Korrelationen effektiv durch die Wechselwirkung zwischen Atomen und zwei-atomigen Molekülen realisiert werden können.

Schließlich legen wir dar, wie das Phänomen der **Spin-Ladungstrennung** mithilfe von stark wechselwirkenden Bosonen in einer räumlichen Dimension beobachtet werden könnte. Für eine Mixtur aus Bosonen mit zwei Isospin-Freiheitsgraden bestimmen wir die effektive Masse einer elementaren Spin-Anregung, die durch den Bethe Ansatz exakt berechnet werden kann. Für das stark korrelierte Regime beweisen wir, dass die effektive Masse einer einzelnen Spin-Anregung die Gesamtmasse aller Teilchen annimmt. Die Spin-Welle propagiert damit wesentlich langsamer als die Dichte-Welle der Bosonen, was der maximalen Form der Spin-Ladungstrennung entspricht.

Contents

Abstract	1
Zusammenfassung	3
Publications	9
1 Introduction	**11**
2 Dynamical creation of a supersolid in bosonic mixtures	**21**
2.1 Introduction	22
2.2 System setup	24
2.3 Equilibrium phase diagram	25
2.4 Out-of-equilibrium preparation of the supersolid	27
2.5 Physical mechanism	31
2.6 Experimental realization	33
2.7 Numerical details	33
2.8 Time evolution of the initial trimer-crystal state	34
2.9 Comparison of the asymptotic state with thermal states	37
2.10 Numerical results of long-time evolutions	40
2.11 Conclusions	41
3 Dynamical creation of bosonic Cooper-like pairs	**45**
3.1 Introduction	45
3.2 Physical sytem	48

3.3	Conservation of pairing	49
3.4	Characterisation of the evolved pair wavefunction	50
3.5	Pair correlations	51
3.6	Numerical details	53
3.7	Cooper-like pairs of bosons	55
3.8	Measuring pair correlations	56
3.9	Conclusions	57

4 Pfaffian-like ground state for 3-body-hard-core bosons — 61

4.1	Introduction	61
4.2	System setup	63
4.3	Ansatz wavefunction for the ground state	64
4.4	Characterisation of the Ansatz wavefunction	66
4.5	Numerical details	67
4.6	Quality of the Ansatz	69
4.7	Experimental proposal	71
4.8	Conclusions	73

5 Spin-charge separation in a one-dimensional spinor Bose gas — 77

5.1	Introduction	77
5.2	System setup	78
5.3	Bethe Ansatz solution	80
5.4	Strong coupling regime	82
5.5	Weak coupling regime	83
5.6	Numerical confirmation	85
5.7	Hydrodynamical approach	86
5.8	Conclusions	87

6 Anyons in one-dimensional optical lattices — 91

6.1	Introduction	91
6.2	Anyon statistics	94

6.3	Fractional Jordan-Wigner mapping	94
6.4	Anyons mapped onto bosons	95
6.5	Restoring left-right symmetry	96
6.6	Experimental proposal	96
6.7	Conclusions	98

Acknowledgements   **105**

Publications

1. **Dynamical creation of a supersolid in asymmetric mixtures of bosons**
   Tassilo Keilmann, J. Ignacio Cirac, and Tommaso Roscilde
   Phys. Rev. Lett. **102**, 255304 (2009).
   See Chapter 2 and Appendix A.

2. **Dynamical creation of bosonic Cooper-like pairs**
   Tassilo Keilmann and Juan José Garcia-Ripoll
   Phys. Rev. Lett. **100**, 110406 (2008).
   See Chapter 3 and Appendix A.

3. **Pfaffian-like ground state for 3-body-hard-core bosons in 1D lattices**
   Belén Paredes, Tassilo Keilmann, and J. Ignacio Cirac
   Phys. Rev. A **75**, 053611 (2007).
   See Chapter 4.

4. **Spin waves in a one-dimensional spinor Bose gas**
   Jean-Noël Fuchs, Dimitri Gangardt, Tassilo Keilmann, and Gora Shlyapnikov
   Phys. Rev. Lett. **95**, 150402 (2005).
   See Chapter 5.

5. **Anyons in 1D optical lattices**
   Tassilo Keilmann and Marco Roncaglia
   In preparation.
   See Chapter 6.

Chapter 1

Introduction

The first realization of Bose-Einstein condensates (BEC) in 1995 [1, 2] opened up new pathways in ultracold atomic physics and provided unique opportunities to explore quantum phenomena associated with weak interactions. Many of the early experiments on BECs can indeed be well explained by mean field theories where interactions do not play a dominant role.

Nowadays, the new challenge on the theoretical side is the strongly interacting and highly correlated regime. Interatomic interactions can be enhanced by tuning the magnetic field across a Feshbach resonance [3], at which the atom-atom scattering length diverges. In a series of remarkable experiments with fermionic atoms this method has been used to observe the crossover from a BEC of molecules to the BCS regime, in which Cooper pairs are formed [4, 5, 6, 7, 8].

An alternative way of attaining the strongly-correlated regime is to load and trap ultracold atoms in an optical lattice potential [9, 10]. By increasing the intensity of the lattice laser beams one can decrease the kinetic energy of the atoms until the interactions dominate the dynamics. Employing this technique, Greiner *et al.* [11] first observed the quantum phase transition from a superfluid to a Mott insulating state of neutral atoms in 2002. In the following years, several groups succeeded in loading bosonic or fermionic atoms into optical lattices and reaching the strongly correlated regime [11, 12, 13, 14, 15, 16, 17, 18, 19]. The optical lattice setup constitutes one of the very few hallmark quantum systems that can be controlled and manipulated on the single quantum level, while at the same time avoiding unwanted interaction with the environment causing decoherence. In addition, optical lattices can be engineered in many different ways to open up new, desirable quantum playgrounds adapted to the needs of the modern physicist. For example, interactions can be tuned from the repulsive to the attractive regime, again by using Feshbach resonances. One can engineer lattices with different geometries, address several internal states of the trapped atoms, or mix fermions with bosons. For its high degree of control and flexibility, it has been proposed to exploit optical lattices to simulate the quantum dynamics of various kinds of Hubbard Hamiltonians [20, 21, 22, 23, 24, 25, 26, 27, 28]. This may help more profoundly to understand the strong correlation effects that have been observed or predicted in solid-state systems. For instance, the study of fermions with repulsive interactions in two dimensions might potentially shed light upon the origin of high temperature superconductivity. Also the physical implementation of a "Feynman quantum simulator" has been put forward [29]. In summary, optical lattices provide a wealth of unique tools to create, study and use the quantum phenomena derived from strongly interacting atoms.

This thesis is devoted to exploit strong correlations among ultracold atoms in order to ultimately create novel, exotic quantum states. With this goal in mind, we will present five different projects in the subsequent chapters.

In Chapter 2 we show how the long-sought *supersolid state* can be created by using current experiments on optical lattices. Supersolids – quantum hybrids exhibiting both superflow and solidity – have been envisioned in 1970 by A. Leggett and G. V. Chester [30, 31]. However, its experimental observation remains elusive. The quest for supersolidity has been strongly revitalized by recent experiments showing possible evidence for a non-zero superfluid fraction present in solid ^{4}He [32]. Yet, several theoretical results appear to rule out the presence of condensation in the pure solid phase of ^{4}He, and various experiments show indeed a strong dependence of the superfluid fraction on extrinsic effects, such as ^{3}He impurities and dislocations. While the experimental findings on bulk ^{4}He remain controversial, optical lattice setups offer the advantages of high sample purity and experimental control to directly pin down a supersolid state via standard measurement techniques.

In Chapter 2, we demonstrate theoretically a *new route to supersolidity*. The key to supersolidity in our setup is a novel non-equilibrium memory effect. By quenching a quantum molecular crystal out of equilibrium, a Bose condensation peak emerges while, surprisingly, the initial solid order is preserved. This memory effect engineers the elusive supersolid state as a quantum superposition between superflow and solidity. We propose that the same principle could be applied to create other exotic forms of excited quantum matter – thus stimulating new directions in the challenging field of quantum state engineering.

In contrast to other theoretical proposals, our model requires only *local interactions*. Longer-range interactions on the Hamiltonian level, which are usually a prerequisite for crystalline order in a supersolid, are here not necessary. On the contrary, effective long-range interactions are created intrinsically by the mass-imbalance of two bosonic species, which arrange in a crystal of trimer molecules. In view of this, our setup is perfectly accessible with current technology, clearing the path to the first experimental observation of supersolidity.

It is widely believed that supersolidity can only appear in quantum crystals with imperfections, where impurities flow coherently through the crystal and build up the condensate fraction. In contrast, we show that such impurities are not necessary and that supersolidity can dynamically occur in a *perfect* quantum crystal.

In Chapter 3 we propose a method to create *Cooper pairs of bosonic atoms* in an optical lattice. Historically, Cooper pairs consist of two fermions with opposite spin and momentum [33]. In our proposal however, we show a way to create a novel state of Cooper-like paired *bosons* – where pairs are macroscopical in size – an effect that has not been observed before in ultracold bosonic atom physics.

The most salient features of this state are that the wavefunction of each pair is a Bell state and that the pair size spans half the lattice, similar to fermionic Cooper pairs.

This bosonic state can be created by a dynamical process that involves crossing a quantum phase transition and which is supported by the symmetries of the physical system. We characterize the final state by means of a measurable two-particle correlator that detects both the presence of the pairs and their size.

In this work, we explore pairing as a property of states, rather than involving energetic considerations, and show that entanglement can be used as a resource to engineer such states. Indeed we demonstrate that a binding energy is not required to support a many-body paired state, but that appropriate symmetries can preserve the pair correlations even through a violent, diabatic evolution.

On the topic of excited many-body states, we emphasize our finding that symmetries and entanglement can be used to create a highly excited non-stationary state with a macroscopic number of pairs. We have thus freed pairing from the constraints of ground-state physics and established it as a new concept in non-equilibrium dynamics. We regard the dynamical process that leads to the pairs by itself as an interesting feature of our setup.

Entanglement is a key concept in this work. First of all, it is our resource for state engineering and the origin of the pair correlation. Second, within our work entanglement acquires an intuitively simple picture related to distributed pairs. Our setup can thus be used as a tool to study the distribution of entanglement, both from the theoretical and the experimental perspectives. In particular, we raise questions about the amount of many-body entanglement that can be achieved by this procedure, the inherent limitations of using fermions or bosons, and whether there are better protocols to achieve extensive correlations in atomic systems.

In Chapter 4 we propose a way to realize the so-called Pfaffian ground state with high fidelity in one-dimensional optical lattices. The elusive Pfaffian state [34] is known to host *non-Abelian anyons* as elementary excitations. Abelian anyons [35] are by definition neither bosons nor fermions but show fractional quantum statistics, multiplying the many-body wavefunction by a fractional, scalar phase factor upon interchange of two such anyons.

Furthermore, non-Abelian anyons [36] exhibit an even more exotic statistical behavior: when two different exchanges are performed consecutively among identical non-Abelian anyons, the final state of the system will depend on the order in which the two exchanges were made. Non-Abelian anyons appeared first in the context of the fractional quantum Hall effect (FQHE) [36], as elementary excitations of exotic states such as the Pfaffian state [34], which is the exact ground state of quantum Hall Hamiltonians with 3-body contact interactions.

In our work, we propose a Pfaffian-like Ansatz for the ground state of bosons subject to 3-body infinite repulsive interactions in a one-dimensional optical lattice. Our Ansatz consists of the symmetrization over all possible ways of distributing the particles in two identical Tonks-Girardeau gases [12]. We support the quality of our Ansatz with numerical calculations and propose an experimental setup based on mixtures of bosonic atoms and molecules in one-dimensional optical lattices, in which this Pfaffian-like state could be realized. Our findings may pave the way to create non-Abelian anyons in one-dimensional systems.

In one-dimensional strongly correlated electron systems, theory predicts that collective excitations of electrons produce, instead of the quasiparticles in ordinary Fermi liquids, two new particles known as "spinons" and "holons" [37]. Unlike ordinary quasiparticles, these particles surprisingly do not carry the spin and charge information of electrons together. Instead, they carry spin and charge information separately and independently. This novel and exotic phenomenon was predicted theoretically decades ago and is commonly known as spin-charge separation.

In Chapter 5, we present a bosonic system where *spin-charge separation* can be realistically maximized: the spin waves (spinons) are shown to be much slower then the charge waves. For this purpose, we consider a two-component (isospin-1/2) Bose gas in a one-dimensional continuous system. The bosons are subject to a spin-independent, repulsive δ-function interaction. We derive exact results for elementary *spin excitations* above the polarized ground state by the Bethe Ansatz method.

We show that – in addition to phonons – the system features spin waves with a quadratic dispersion. Furthermore, we compute analytically and numerically the effective mass of the spin wave and show that the spin transport is greatly suppressed in the strong coupling regime. Remarkably, the effective mass of the elementary spin excitation reaches the *total mass* of all bosons in the system in the strong coupling limit. In this regime, the bosons are impenetrable and therefore a spin excited boson can move on the one-dimensional ring only if all other bosons move as well.

In this work we have thus found extremely slow spin dynamics in the strongly correlated regime, originating from a very large effective mass of spin waves. In an experiment with ultracold bosons, this effect should show up as a spectacular *isospin-density separation*: an initial wave packet splits into a fast acoustic (charge) wave traveling at the Fermi velocity and an extremely slow spin wave. One can even think of "freezing" the spin transport, which in experiments with two-component one-dimensional Bose gases will correspond to freezing relative oscillations of the two components, maximizing the spin-charge separation.

Finally, in Chapter 6 we propose a way to realize a *gas of Abelian anyons* [35] in an optical lattice. We establish analytically an exact mapping between anyons and bosons in one dimension, via a generalized Jordan-Wigner transformation. We will show that anyons trapped in a one-dimensional lattice are equivalent to – and can be realized by – ordinary bosons with conditional hopping amplitudes.

This work is still unfinished. At this stage, we are presenting the analytical heart of the project, which proves the mapping between anyons and "conditional-hopping bosons" on a lattice. Furthermore, we give an outlook concerning the realization of this specific bosonic model, discussing an experimental scheme involving laser-assisted, state-dependent tunneling, by which the anyon statistics angle θ can be directly controlled.

The experimental implementation of the bosonic Hamiltonian proposed in Chapter 6 would directly give rise to the realization of the long-sought many-particle state of *anyons*.

References

[1] M. H. Anderson, J. R. Ensher, M. R. Matthews, C. E. Wieman, and E. A. Cornell, Science **269**, 198201 (1995).

[2] K. B. Davis, M.-O. Mewes, M. R. Andrews, N. J. van Druten, D. S. Durfee, D. M. Kurn, and W. Ketterle, Phys. Rev. Lett. **75**, 3969 (1995).

[3] H. Feshbach, Ann. Phys. **5**, 357 (1958).

[4] C. A. Regal, M. Greiner, and D. S. Jin, Phys. Rev. Lett. **92**, 040403 (2004).

[5] M. W. Zwierlein, C. A. Stan, C. H. Schunck, S. M. F. Raupach, A. J. Kerman, and W. Ketterle, Phys. Rev. Lett. **92**, 120403 (2004).

[6] M. Bartenstein, A. Altmeyer, S. Riedl, S. Jochim, C. Chin, J. H. Denschlag, and R. Grimm, Phys. Rev. Lett. **92**, 120401 (2004).

[7] T. Bourdel, L. Khaykovich, J. Cubizolles, J. Zhang, F. Chevy, M. Teichmann, L. Tarruell, S. J. J. M. F. Kokkelmans, and C. Salomon, Phys. Rev. Lett. **93**, 050401 (2004).

[8] J. Kinast, S. L. Hemmer, M. E. Gehm, A. Turlapov, and J. E. Thomas, Phys. Rev. Lett. **92**, 150402 (2004).

[9] P. Verkerk *et al.*, Phys. Rev. Lett. **68**, 3861 (1992)

[10] P. S. Jessen *et al.*, Phys. Rev. Lett. **69**, 49 (1992)

[11] M. Greiner, O. Mandel, T. Esslinger, T. W. Hänsch, and I. Bloch, Nature **415**, 39 (2002).

[12] B. Paredes *et al.*, Nature **429**, 277 (2004).

[13] H. Moritz, T. Stöferle, M. Kohl, and T. Esslinger, Phys. Rev. Lett. **91**, 250402 (2003).

[14] T. Stöferle, H. Moritz, C. Schori, M. Köhl, and T. Esslinger, Phys. Rev. Lett. **92**, 130403 (2004).

[15] M. Köhl, H. Moritz, T. Stöferle, K. Günter, and T. Esslinger, Phys. Rev. Lett. **94**, 080403 (2005).

[16] B. Laburthe Tolra, K. M. OHara, J. H. Huckans, W. D. Phillips, S. L. Rolston, and J. V. Porto, Phys. Rev. Lett. **92**, 190401 (2004).

[17] K. Xu, Y. Liu, J. R. Abo-Shaeer, T. Mukaiyama, J. K. Chin, D. E. Miller, W. Ketterle, K. M. Jones, and E. Tiesinga, Phys. Rev. A **72**, 043604 (2005).

[18] C. Ryu, X. Du, E. Yesilada, A. M. Dudarev, S. Wan, Q. Niu, and D. Heinzen, arXiv:cond-mat/0508201.

[19] G. Thalhammer, K. Winkler, F. Lang, S. Schmid, R. Grimm, and J. H. Denschlag, Phys. Rev. Lett. **96**, 050402 (2006).

[20] J. I. Cirac and P. Zoller, Science **301**, 176 (2003).

[21] J. I. Cirac and P. Zoller, Phys. Today **57**, 38 (2004).

[22] W. Hofstetter, J. I. Cirac, P. Zoller, E. Demler, and M. Lukin, Phys. Rev. Lett. **89**, 220407 (2002).

[23] J. J. Garcia-Ripoll and J. I. Cirac, New J. Phys. **5**, 76 (2003).

[24] L.-M. Duan, E. Demler, and M. D. Lukin, Phys. Rev. Lett. **91**, 090402 (2003).

[25] J. J. Garcia-Ripoll, M. A. Martin-Delgado, and J. I. Cirac, Phys. Rev. Lett. **93**, 250405 (2004).

[26] L. Santos, M. A. Baranov, J. I. Cirac, H.-U. Everts, H. Fehrmann, and M. Lewenstein, Phys. Rev. Lett. **93**, 030601 (2004).

[27] H. P. Büchler, M. Hermele, S. D. Huber, M. P. A. Fisher, and P. Zoller, Phys. Rev. Lett. **95**, 040402 (2005).

[28] A. Micheli, G. K. Brennen, and P. Zoller, Nature Physics **2**, 341-347 (2006).

[29] E. Jané, G. Vidal, W. Dür, P. Zoller, and J. I. Cirac, Quant. Inf. Comp. **3**, 15 (2003).

[30] A. J. Leggett, Phys. Rev. Lett. **25**, 1543 (1970).

[31] G. V. Chester, Phys. Rev. A **2**, 256 (1970).

[32] E. Kim, M. H. W. Chan, Nature **427**, 225-227 (2004).

[33] J. Bardeen, L. N. Cooper, and J. R. Schrieffer, Phys. Rev. **108**, 1175 - 1204 (1957).

[34] M. Greiter, X. G. Wen, and F. Wilczek, Nucl. Phys. B **374**, 567 (1992); C. Nayak and F. Wilczek, Nucl. Phys. B **479**, 529 (1996).

[35] G. S. Canright and S. M. Girvin, Sciene **247**, 1197 (1990).

[36] G. Moore and N. Read, Nucl. Phys. B **360**, 362 (1991).

[37] E. H. Lieb, F. Y. Wu, Phys. Rev. Lett. **20**, 1445 (1968).

Chapter 2

Dynamical creation of a supersolid in bosonic mixtures

Supersolidity – the simultaneous appearance of spontaneous solid and superfluid order [1, 2] – is a long-sought quantum phase in many-body physics. A recent, vibrant debate on its possible realization in current experiments on quantum crystals [3, 4, 5] has posed the fundamental question on whether supersolidity can be an intrinsic property of a perfect quantum crystal, or whether it necessitates extrinsic agents such as imperfections. Here we show theoretically that a supersolid can appear in a perfect one-dimensional crystal – without the requirement of doping – created by an attractive mixture of mass-imbalanced hardcore bosons in an optical lattice. Starting from a "molecular" quantum crystal, supersolidity is induced dynamically as an out-of-equilibrium state. When neighboring molecular wavefunctions overlap, both bosonic species simultaneously exhibit quasi-condensation and long-range solid order, which is stabilized by their mass imbalance. Our model can be realized in present experiments with bosonic mixtures that feature simple on-site interactions, clearing the path to the first observation of supersolidity.

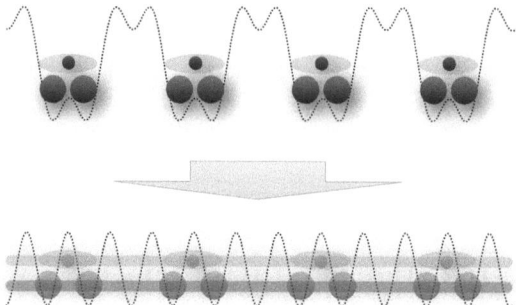

Figure 2.1: **Schematic of the quantum quench leading to supersolidity.** A product state of bosonic trimers is the initial state of the evolution (larger symbols represent the ↓-bosons); switching off one of the superlattice components leads to a supersolid state in which the particles delocalize into a (quasi-)condensate while maintaining the original solid pattern without imperfections.

2.1 Introduction

The intriguing possibility of creating a quantum hybrid exhibiting both superflow and solidity has been envisioned long ago [1, 2]. However, its experimental observation remains elusive. The quest for supersolidity has been strongly revitalized by recent experiments showing possible evidence for a non-zero superfluid fraction present in solid ^{4}He [3]. Yet, several theoretical results [6] appear to rule out the presence of condensation in the pure solid phase of ^{4}He, and various experiments [7] show indeed a strong dependence of the superfluid fraction on extrinsic effects, such as ^{3}He impurities and dislocations. While the experimental findings on bulk ^{4}He remain controversial, optical lattice setups [8] offer the advantages of high sample purity and experimental control to directly pin down a supersolid state via standard measurement techniques. A variety of lattice boson models with strong finite-range interactions has

2.1 Introduction

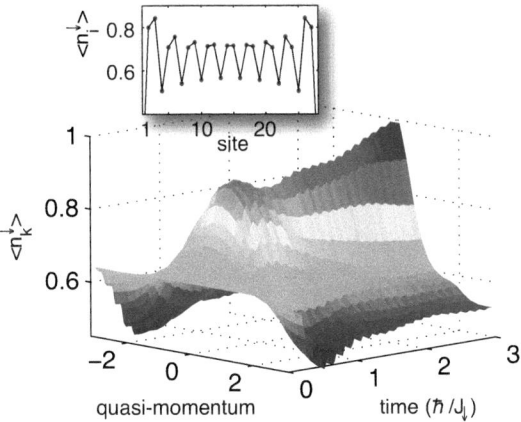

Figure 2.2: **Dynamical onset of supersolidity by quantum quenching a mixture of light and heavy bosons.** Momentum profile of the $\downarrow$-bosons, $\langle n_k^\downarrow \rangle$ vs. time in units of hopping events $\hbar/J_\downarrow$. A quasi-condensate peak develops rapidly. Inset: Density distribution $\langle n_i^\downarrow \rangle$ averaged over the last third of the evolution time, showing that crystalline order is conserved in the system. The simulation parameters are $L = 28$, $N_\downarrow = 18$, $N_\uparrow = 9$, $J_\downarrow/J_\uparrow = 0.1, U/J_\uparrow = 3.0$.

been recently shown to display crystalline order and supersolidity upon doping the crystal state away from commensurate filling [6, 9]; yet sizable interactions with a finite range are not available in current cold-atom experiments. Such interactions can be in principle obtained effectively by adding a second atomic species of fermions [10, 11], which, however, does not participate in the condensate state, in a way similar to the nuclei forming the lattice of a superconductor without participating in the condensate of electron pairs.

Here we demonstrate theoretically a new route to supersolidity, realized as the out-of-equilibrium state of a realistic lattice-boson model after a so-called "quantum quench" (a sudden change in the Hamiltonian). The equilibrium Hamiltonian of the model before the quench realizes a "molecular crystal" phase characterized by the crystallization of atomic trimers made of two mass-imbalanced bosonic species. Starting from a solid of tightly-bound trimers and suddenly changing the system Hamiltonian, the evolution induces broadening and overlap of neighboring molecular wavefunctions leading to quasi-condensation of all atomic species, while crystalline order is maintained, see Figs. 2.1 and 2.2. Our model requires only local on-site interactions as currently featured by neutral cold atoms, which make the observation of a supersolid state a realistic and viable goal.

2.2 System setup

We consider two bosonic species ($\sigma = \uparrow, \downarrow$) tightly confined in two transverse spatial dimensions and loaded in an optical lattice potential in the third dimension. In the limit of a deep optical lattice, the dynamics of the atoms can be described by a model of lattice hardcore bosons in one dimension [10, 13]

$$\mathcal{H} = -\sum_{i,\sigma} J_\sigma \left(b_{i,\sigma}^\dagger b_{i+1,\sigma} + \text{h.c.} \right) - U \sum_i n_{i,\uparrow} n_{i,\downarrow}. \tag{2.1}$$

Here the operator $b_{i\sigma}^\dagger$ ($b_{i\sigma}$) creates (annihilates) a hardcore boson of species σ on site i of a chain of length L, and it obeys the on-site anticommutation relations $\{b_{i\sigma}, b_{i\sigma}^\dagger\} = 1$. $n_{i\sigma} \equiv b_{i\sigma}^\dagger b_{i\sigma}$ is the number operator. Throughout this work we restrict ourselves to the case of attractive on-site interactions $U > 0$ and to the case of mass imbalance, $J_\uparrow > J_\downarrow$. Moreover we fix the lattice fillings of the two species to $n_\uparrow = 1/3$ and $n_\downarrow = 2/3$.

In the extreme limit of mass imbalance, $J_\downarrow = 0$, equation (2.1) reduces to the well-known Falicov–Kimball model of mobile particles in a potential created by static impurities [15]. For the considered filling it can be shown via exact diagonalization that, at sufficiently low attraction $U/J_\uparrow \leq 2.3$, the ground state realizes a crystal of *trimers* formed by two $\downarrow$-bosons "glued" together by an $\uparrow$-boson in an atomic analogue of a covalent bond (see Fig. 2.1 for a scheme of the spatial arrangement). The trimer crystal is protected by a finite energy gap against dislocations of the ↓-bosons, and hence it is expected to survive the presence of a small hopping $J_\downarrow$.

2.3 Equilibrium phase diagram

Indeed extensive quantum Monte Carlo simulations (see numerical details in section 2.7) show that the ground state phase diagram features an extended trimer crystal phase (Fig. 2.3). For $U/J_\uparrow > 2.3$, and over a large region of $J_\downarrow/J_\uparrow$ ratios the ground state shows instead the progressive merger of the trimers into hexamers, dodecamers, and finally into a fully collapsed phase with phase separation of the system into particle-rich and particle-free regions.

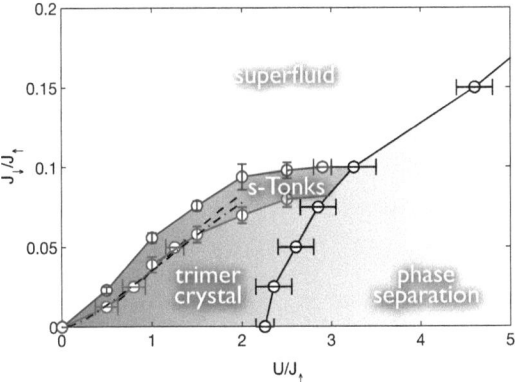

Figure 2.3: **Equilibrium phase diagram (ground state).** The dash-dotted line represents the points where the hopping of the ↓-bosons, $J_\downarrow$, overcomes the energy gap to crystal dislocations, giving rise to the solid/super-Tonks (s-Tonks) transition. The dashed line marks the points where a single-trimer wavefunction spreads over 2.8 sites. In the super-Tonks phase, quasi-solidity and superfluidity coexist.

For $U/J_\uparrow \leq 2.3$, increasing the $J_\downarrow/J_\uparrow$ ratio allows to continuously tune the zero-point quantum fluctuations of the ↓-atoms in the trimer crystal and to increase the effective size of the trimers, whose wavefunctions start to overlap. We find that, when trimers spread over a critical size of ≈ 2.8 lattice sites, they start exchanging atoms and the quantum melting of the crystal is realized. The melting point is also consistent with the point at which the hopping $J_\downarrow$ overcomes the energy gap to dislocations (dash-dotted line in Fig. 2.3). The resulting phase after quantum melting is a one-dimensional superfluid for both atomic species: in this phase quasi-condensation appears, in the form of power-law decaying phase correlations

$$\langle b^\dagger_{i,\sigma} b_{j,\sigma}\rangle \propto |r_i - r_j|^{-\alpha_\sigma}, \qquad (2.2)$$

which is the strongest form of off-diagonal correlations possible in interacting one-dimensional quantum models [18]. Yet in the superfluid phase strong power-law density correlations survive,

$$\langle n_{i,\sigma} n_{j,\sigma}\rangle \propto \cos(q_{\text{tr}}(r_i - r_j)) |r_i - r_j|^{-\beta_\sigma}, \qquad (2.3)$$

exhibiting oscillations at the trimer-crystal wavevector $q_{\text{tr}} = 2\pi/3$. Such correlations stand as remnants of the solid phase, and in a narrow parameter region they even lead to a *divergent* peak in the density structure factor, $S_\sigma(q_{\text{tr}}) \propto L^{\beta_\sigma}$ with $0 < \beta_\sigma < 1$, where

$$S_\sigma(q) = \frac{1}{L} \sum_{ij} e^{iq(r_i - r_j)} \langle n_{i,\sigma} n_{j,\sigma}\rangle. \qquad (2.4)$$

This phase, termed "super-Tonks" phase in the literature on one-dimensional quantum systems [16], is a form of quasi-supersolid, in which one-dimensional superfluidity coexists with quasi-solid order. (Notice that true solidity corresponds to $\beta_\sigma = 1$.)

2.4 Out-of-equilibrium preparation of the supersolid

The strong competition between solid order and superfluidity in the ground-state properties of this model suggests the intriguing possibility that true supersolidity might appear by perturbing the system out of the above equilibrium state. In particular we investigate the Hamiltonian evolution of the system after its state is prepared out of equilibrium in a perfect trimer crystal. The initial state is a simple factorized state of perfect trimers (see Fig. 2.1):

$$|\Psi_0\rangle = \bigotimes_{n=1}^{L/3} |\Phi_{\text{tr}}^{(3n-1)}\rangle \tag{2.5}$$

where the trimer wavefunction reads

$$|\Phi_{\text{tr}}^{(i)}\rangle = \frac{1}{\sqrt{2}} b_{i\downarrow}^\dagger b_{i+1\downarrow}^\dagger (b_{i\uparrow}^\dagger + b_{i+1\uparrow}^\dagger)|\text{vac}\rangle. \tag{2.6}$$

This state can be realized with the current technology of optical superlattices [3], by applying a strong second standing wave component $V_{x_2} \cos^2[(k/3)x + \pi/2]$ to the primary wave, $V_{x_1} \cos^2(kx)$, creating the optical lattice along the x direction of the chains. This superlattice potential has the structure of a succession of double wells separated by an intermediate, high-energy site. Hence tunneling out of the double wells is strongly suppressed, stabilizing the factorized state, equation (2.5).

After preparation of the system in the initial state, the second component of the superlattice potential is suddenly switched off ($V_{x_2} \to 0$) and the state is let to evolve with the Hamiltonian corresponding to different parameter sets ($U/J_\uparrow, J_\downarrow/J_\uparrow$). The successive time evolution over a short time interval $[0, \tau]$ with $\tau = 3\hbar/J_\downarrow$ is computed using the Matrix-Product-States (MPS) algorithm on a one-dimensional lattice with up to 28 sites and open boundary conditions [14, 19], see also Numerical details. We characterize the evolved state by averaging the most significant observables over the last portion of the time evolution $\tau/3$.

We find three fundamentally different evolved states, whose extent in parameter space is shown on the non-equilibrium phase diagram of Fig. 2.4:

Firstly, we find a superfluid phase, in which the initial crystal structure is completely

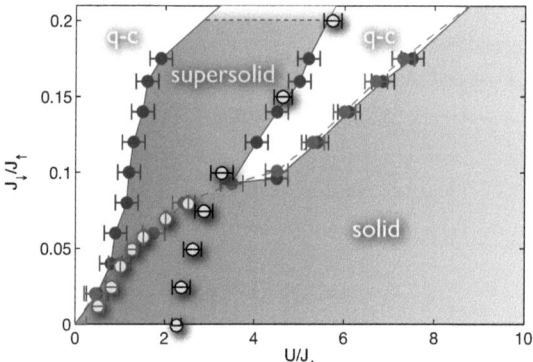

Figure 2.4: **Out-of-equilibrium phase diagram.** An extended supersolid phase exists in the transient state attained after the quantum quench. In this phase true solidity and quasi-condensation coexist. Blue symbols delimit the boundaries of the solid phase, red symbols mark the lower boundary for the quasi-condensed (q-c) phase. The overlap of both phases (blue shaded region) is identified as the supersolid phase. The yellow-filled symbols correspond to equilibrium data points. The lower boundary of the superfluid/super-Tonks region of the equilibrium phase diagram is seen to coincide with the lower boundary of the supersolid region out of equilibrium.

melted by the Hamiltonian evolution, and coherence builds up in the system leading to quasi-condensation out-of-equilibrium, namely to the appearance of a (sub-linearly) diverging peak in the momentum distribution

$$\langle n_k^\sigma \rangle = \frac{1}{L} \sum_{ij} e^{ik(r_i - r_j)} \langle b_{i,\sigma}^\dagger b_{j,\sigma} \rangle \qquad (2.7)$$

at zero quasimomentum, $\langle n_{k=0}^\sigma \rangle \propto L^{\alpha_\sigma}$ with $0 < \alpha_\sigma < 1$. Despite the short time evolution, quasi-condensation of the slow $\downarrow$-bosons is probably assisted by their interaction with the faster $\uparrow$-bosons, and is observed to occur for all system sizes considered. Secondly, we find a solid phase, in which the long-range crystalline phase of the ini-

2.4 Out-of-equilibrium preparation of the supersolid

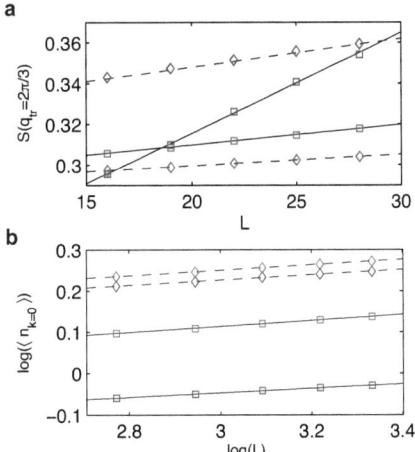

Figure 2.5: **Coexistence of solid order and quasi-condensation in the supersolid phase.** (a) The structure factor peak $S(q_{tr} = 2\pi/3)$ scales linearly with system size L, demonstrating solid order for both bosonic species. (b) The density peak in momentum space $\langle n^{\downarrow}_{k=0} \rangle$ is plotted vs. L on a log-log scale, showing algebraic scaling and thus quasi-condensation. Boxes (diamonds) stand for particle species $\downarrow$ ($\uparrow$), respectively. The data represented by blue boxes in part (a) is offset by -0.2 for better visibility. Parameters: $J_\downarrow/J_\uparrow = 0.1, U/J_\uparrow = 3.0$ (blue symbols) and $J_\downarrow/J_\uparrow = 0.15, U/J_\uparrow = 2.5$ (red symbols).

tial state is preserved, as shown by the structure factor which has a linearly diverging peak at the trimer-crystal wavevector $S(q_{tr}) \propto L$.

Thirdly, an extended supersolid phase emerges, with perfect coexistence of the two above forms of order for both atomic species. This is demonstrated in Fig. 2.5 via the finite-size scaling of the peaks in the momentum distribution and in the density structure factor. In this phase, which has no equilibrium counterpart, the Hamiltonian evolution leads to the delocalization of a significant fraction of $\uparrow$- and $\downarrow$-bosons over the entire system size. Consequently quasi-long-range coherence builds up and

2. Dynamical creation of a supersolid in bosonic mixtures

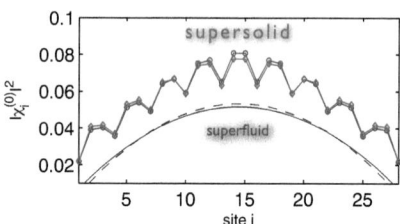

Figure 2.6: **Snapshot of a supersolid.** Square modulus of the natural orbital $\chi_i^{(0)}$ corresponding to the largest eigenvalue of the OBDM, calculated at final time τ. In the supersolid regime (blue/red symbols for $\downarrow/\uparrow$ bosons), the natural orbital shows the characteristic crystalline order. This pattern is washed out in the purely quasi-condensed regime (dashed/solid curves for $\downarrow/\uparrow$). The supersolid data is offset by $+0.02$ for the sake of visibility. Parameters: $J_\downarrow/J_\uparrow = 0.1$ (supersolid), $J_\downarrow/J_\uparrow = 0.8$ (quasi-condensed), $U/J_\uparrow = 3.0$, $N_\downarrow = 18$, $N_\uparrow = 9$, $L = 28$.

the momentum distribution, which is completely flat in the initial localized trimer-crystal state, acquires a pronounced peak at zero quasi-momentum $k = 0$, as shown in Fig. 2.2. Yet the quasi-condensation order parameter $\chi_i^{(0)}$, namely the natural orbital of the one-body density matrix (OBDM) $\langle b_{i,\sigma}^\dagger b_{j,\sigma} \rangle$ corresponding to the largest eigenvalue and hosting the condensed particles, is spatially modulated (cf. Fig. 2.6), revealing the persistence of solid order in the quasi-condensate. In addition, solidity can be confirmed by direct inspection of the real-space density $\langle n_{i\sigma} \rangle$ (cf. inset of Fig. 2.2). Going from the boundaries towards the center, the density profiles of both species are modulated by the crystal structure, and the modulation amplitudes saturate at constants which turn out to be independent of the system size.

To gain further insight into the mechanism underlying the stabilization of a commensurate two-species supersolid via out-of-equilibrium time evolution, we finally compare the equilibrium phase diagram with the non-equilibrium one. Fig. 2.4 shows that the superfluid/solid and superfluid/phase-separation boundaries at equilibrium overlap with the threshold of formation of the supersolid out of equilibrium upon increasing $J_\downarrow/J_\uparrow$. This means that a quantum quench of the system Hamiltonian to the parameter range corresponding to a superfluid equilibrium ground state is a necessary condition for supersolidity to dynamically set in.

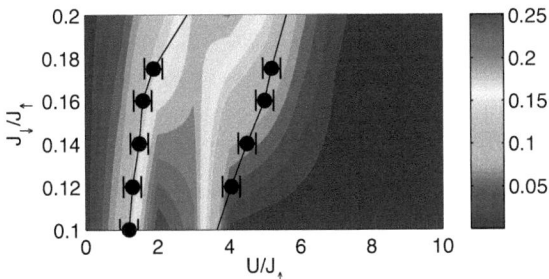

Figure 2.7: **Overlap of the equilibrium ground state with the initial trimer-crystal state.** The overlap $|c_0|^2$ (contour plot) agrees well with the boundaries of the non-equilibrium supersolid phase (black symbols, cf. Fig. 2.4). This suggests a superfluid ground state as a necessary condition for supersolidity to dynamically set in. The overlap $|c_0|^2$ has been calculated via exact diagonalization on a $L = 10$ chain containing three trimers.

2.5 Physical mechanism

The key to the dynamical emergence of a quasi-condensate fraction in the supersolid phase is that the initial trimer-crystal state, equation (2.5), has a significant overlap with the superfluid ground state of the final Hamiltonian after the quantum quench. As shown in Fig. 2.7 the ground-state overlap $|c_0|^2$ remains sizable over an extended parameter range. This is intimately connected with the strong density–density correlations present in the equilibrium superfluid phase, as shown e.g. by the appearance of a region with super-Tonks behavior. The excellent agreement between the region featuring supersolidity and the region with most pronounced overlap $|c_0|^2$ suggests the following mechanism: the Hamiltonian evolution following the quantum quench dynamically selects the ground-state component as the one giving the dominant contribution to (quasi-)long-range coherence. In essence, while the quantum melting phase transition occurring at equilibrium leads to a dichotomy between solid and superfluid order, the out-of-equilibrium preparation can coherently admix the excited crystalline states with the superfluid ground state without disrupting their respective forms of order (see section 2.8). It is tempting to think that a similar preparation scheme of supersolid states can work in other systems displaying solid–superfluid phase boundaries at equilibrium.

The supersolid transient state discussed in this work may ultimately lead to a supersolid steady state for longer evolution times $\tau \gg 3\hbar/J_\perp$ which, however, are not accessible with present numerical methods. An intriguing question arises when considering the asymptotic time limit: Does supersolidity survive or is long-range order ultimately destroyed by thermalisation? Recent numerical studies of other strongly correlated one-dimensional quantum systems reveal a failure of thermalisation [22, 23]. We have considered the asymptotic time limit using exact diagonalization for a small system (see section 2.9). These exact results suggest that supersolidity persists and the system does not converge to an equilibrium thermal state (in fact even thermalisation in the microcanonical ensemble [21] does not seem to occur in our system, see section 2.9). Whether the absence of thermalisation persists when taking the thermodynamic limit remains an open question, that can be answered only by experiments. In view of this, an experimental realization of our proposal could not only be used to create the elusive supersolid state at short times, but also to ultimately test whether thermalisation sets in or not.

2.6 Experimental realization

The observation of the supersolid state prepared via the dynamical scheme proposed in this work is directly accessible to several setups in current optical-lattice experiments. The fundamental requirement to explore the phase diagrams of our model, Figs. 2.3 and 2.4, is the existence of a stable bosonic mixture with mass imbalance and interspecies interactions that can be tuned to the attractive regime via a Feshbach resonance. This requirement is met in spin mixtures of, e.g., ^{87}Rb atoms in different hyperfine states, which acquire a spin-dependent effective mass when loaded in an optical lattice [14], and for which Feshbach resonances have been extensively investigated [24]. Moreover recently discovered Feshbach resonances in ultracold heteronuclear bosonic mixtures (^{87}Rb-^{133}Cs, ^{7}Li-^{87}Rb, ^{41}K-^{87}Rb, ^{39}K-^{87}Rb and others [25], the latter recently loaded in optical lattices [26]) enlarge even further the number of candidate systems to implement the Hamiltonian, equation (2.1). The hardcore-repulsive regime can be easily accessed in deep optical lattices [13]. After preparation of the trimer crystal via an optical superlattice [3], the onset of coherence in the supersolid state, attained after a short hold time corresponding to $\approx$ 2-3 hopping events of the slower particles ($\approx$ 1-10 ms), can be monitored by time-of-flight measurements of the momentum distribution. The rapid onset of coherence allows the experimental detection of supersolidity even before decoherence effects become important. On the other hand, the persistence of the crystalline structure can be probed by resonant Bragg scattering [27]. While experimentally the initial state will be always a mixed one and not the pure state in equation (2.5), we observe that mixedness of the initial state does not disrupt supersolidity in the evolved state.

2.7 Numerical details

The equilibrium phase diagram of Fig. 2.3 has been obtained via quantum Monte Carlo simulations based on the canonical Stochastic Series Expansion algorithm [28, 29]. Simulations have been performed on chains of size $L = 30, ..., 120$ with periodic boundary conditions, at an inverse temperature $\beta J_\downarrow = 2L/3$ ensuring that the obtained data describe the zero-temperature behaviour for both atomic species. The border between the superfluid phases and the solid/phase-separated phases in the phase diagram of Fig. 2.3 has been obtained by analyzing the superfluid fraction, calculated via the winding number of the worldlines in the Monte Carlo simulation. In the Falicov–Kimball limit $J_\downarrow = 0$ we have performed exact diagonalizations for a system of hardcore bosons, mapped onto spinless fermions [18], in an adjustable static potential created by the $\downarrow$-particles.

The out-of-equilibrium phase diagram of Fig. 2.4 and all the data plotted in Figs. 2.2, 2.5 and 2.6 have been obtained with a Matrix-Product-State algorithm for Hamiltonian time evolution [14, 19]. A bond dimension $D = 500$ ensures that the weight of the discarded Hilbert space is $< 10^{-3}$. The evolution time step $dt = 5 \times 10^{-3} \hbar / J_\uparrow$ is chosen so as to make the Trotter error smaller than 10^{-3}. The phase diagram of Fig. 2.4 has been obtained via finite-size scaling on five different system sizes $L = 16, ..., 28$. The scaling behaviour of the observables $S_\sigma(q_{\text{tr}})$ and $\langle n_{k=0}^\sigma \rangle$ has been used to identify the different phases.

2.8 Time evolution of the initial trimer-crystal state

In the following we present exact calculations for a small system which elucidate the special nature of the initial trimer crystal state after the quench, superimposing the superfluid (and quasi-condensed) ground state with selected crystalline excited states. Furthermore, we compare the results for the asymptotic state of the time evolution with thermal states in both the canonical and microcanonical ensembles. Our results indicate that thermalisation may not occur in our system.

We discuss here in more detail the time evolution of the initial trimer-crystal state into a supersolid state. The initial state, equation (2.5), can be decomposed into the eigenstates of the final Hamiltonian $\mathcal{H}|E_a\rangle = E_a|E_a\rangle$ as:

$$|\Psi(t=0)\rangle = \sum_a c_a |E_a\rangle. \tag{2.8}$$

2.8 Time evolution of the initial trimer-crystal state

The time-evolved state is then:

$$|\Psi(t)\rangle = \sum_a c_a e^{-i\omega_a t}|E_a\rangle \tag{2.9}$$

where $\omega_a = E_a/\hbar$.

The expectation value of any operator A can be thus written as

$$\begin{aligned}\langle A\rangle_t &= \sum_a |c_a|^2 \langle E_a|A|E_a\rangle \\ &+ \sum_{a\neq b} 2\,\mathrm{Re}\left[\langle E_a|A|E_b\rangle c_a^* c_b e^{i(\omega_a-\omega_b)t}\right]\end{aligned} \tag{2.10}$$

approaching the "diagonal ensemble" [21] or steady state for a large time $t \to \infty$,

$$\langle A\rangle_\infty = \sum_a |c_a|^2 \langle E_a|A|E_a\rangle. \tag{2.11}$$

We now specify the discussion to the case in which the system is evolved with a quantum Hamiltonian whose ground state is both a superfluid and a quasi-condensate. If the initial trimer-crystal state has a significant overlap with the quasi-condensed ground state, namely if c_0 is not negligible, then one can expect that the phase correlator of the steady state, corresponding to $A \equiv b^\dagger_{i,\sigma} b_{j,\sigma}$, will be dominated by the ground-state contribution, so that (quasi-)long-range order sets in. At the same time, the initial state has by construction a significant projection on excited states $|E_{a>0}\rangle$ with long-range crystalline correlations, provided that these states exist in the Hamiltonian spectrum. Under this assumption, the density-density correlator, corresponding to $A \equiv n_{i,\sigma} n_{j,\sigma}$, will remain long-ranged in the steady state; this fact, combined with (quasi-)long-range phase coherence, gives rise to supersolidity.

Making use of exact diagonalization on a $L = 10$ chain with open boundary conditions, we have systematically investigated the overlap c_0 between the perfect trimer-crystal state and the Hamiltonian ground state for different points in parameter space. The results are shown in Fig. 2.7, and compared with the phase boundaries of the non-equilibrium phase diagram, Fig. 2.4. We observe that the non-equilibrium supersolid phase is in striking correspondence with the parameter region where c_0

is largest, suggesting that the above analysis of the onset of supersolidity is quantitatively correct. Note that the time evolution discussed in the previous sections is restricted to finite times, while we focus here on the asymptotic case $t \to \infty$.

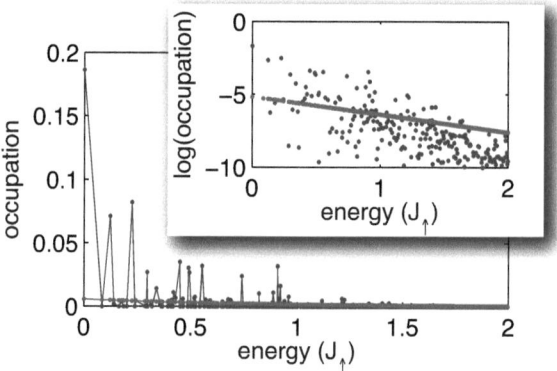

Figure 2.8: **Diagonal vs. thermal probability distributions.** The occupations of the diagonal ($|c_a|^2$ in blue) and canonical ($|d_a|^2$ in green) ensembles are plotted as a function of the eigenstate energies (offset from E_{GS}). Contrary to the thermal, continuous distribution, the trimer-crystal state emphasizes certain eigenstates, while it suppresses others. The (superfluid) ground state contribution present in the trimer-crystal state is enhanced by a factor of ≈ 20 compared with the thermal contribution. Most of the amplified excited states indeed show a crystalline structure with the correct periodicity, or contain density peaks at the right positions to build up the final crystal. Inset: The same distributions on a log-lin scale. The deviation of the diagonal from the thermal ensemble is even better visualized here.

2.9 Comparison of the asymptotic state with thermal states

The diagonal-ensemble expectation value of equation (2.11) is here compared with a thermal average in the canonical ensemble

$$\langle A \rangle_T = \sum_a |d_a|^2 \langle E_a|A|E_a \rangle, \qquad (2.12)$$

with $|d_a|^2 = \exp(-E_a/k_B T)/Z$ the Boltzmann weights, k_B Boltzmann's constant, T the temperature and $Z = \sum_a \exp(-E_a/k_B T)$ the normalizing partition function. In addition, we introduce for comparison the statistical average in the microcanonical ensemble

$$\langle A \rangle_{E_{\text{in}}, dE} = \sum_{E_{\text{in}}-dE < E_a < E_{\text{in}}+dE} 1/N_m \langle E_a|A|E_a \rangle, \qquad (2.13)$$

which averages over eigenstates within an energy window $\pm dE$ around the initial energy

$$E_{\text{in}} = \langle \Psi(t=0)|\mathcal{H}|\Psi(t=0) \rangle. \qquad (2.14)$$

N_m is the number of eigenstates contained in that energy window.

In order to compare the diagonal with the canonical and microcanonical ensembles, we have chosen to exactly diagonalize a system of three trimers ($N_\downarrow = 6, N_\uparrow = 3$) in an open chain of $L = 10$ sites. We present in the following results for the parameter pair ($J_\downarrow = 0.2 J_\uparrow, U = 3 J_\uparrow$), where supersolidity exists according to our non-equilibrium phase diagram, Fig. 2.4. Under these conditions, the ground state energy yields $E_{GS} \simeq -13.2 J_\uparrow$, while the initial trimer-crystal state carries an energy $E_{\text{in}} = -12 J_\uparrow$. In order to determine the correct temperature for the canonical ensemble, we have varied T until the condition $\langle \mathcal{H} \rangle_T = E_{\text{in}}$ was met. This analysis yielded $k_B T \simeq 0.82 J_\uparrow$, which we use henceforth for the comparison with the canonical averages.

Fig. 2.8 compares the diagonal ensemble induced by the initial trimer-crystal state with a thermal, canonical ensemble. The trimer-crystal state has a finite projection on the quasi-condensed ground state as well as on distinct excited states. Further inspection into those excited states shows that their characteristic density profiles

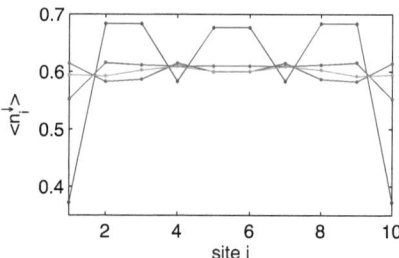

Figure 2.9: **Real-space density $\langle n_i^\downarrow \rangle$ in all three ensembles.** While the diagonal ensemble $\langle n_i^\downarrow \rangle_\infty$ (blue) shows a clear crystalline pattern, this structure is washed out completely in the canonical ensemble $\langle n_i^\downarrow \rangle_{T=0.82 J_\uparrow/k_B}$ (green). Results for the microcanonical ensemble $\langle n_i^\downarrow \rangle_{E_{\text{in}}, dE}$ are shown for energy windows $dE = 0.2 J_\uparrow$ (red) and $dE = 0.6 J_\uparrow$ (cyan). All thermal ensembles deviate strongly from the density structure at time $t \to \infty$ (diagonal ensemble).

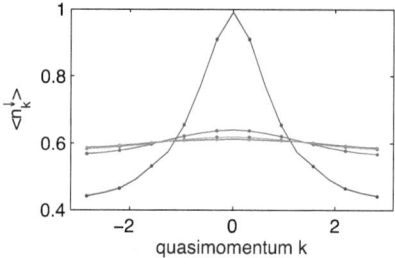

Figure 2.10: **Momentum profile $\langle n_k^\downarrow \rangle$ in all three ensembles.** Due to the significant weight attributed to the ground state, the diagonal ensemble $\langle n_k^\downarrow \rangle_\infty$ (blue) features an enhanced quasi-condensation peak at $k = 0$. This peak is suppressed in all thermal ensembles $\langle n_k^\downarrow \rangle_{T=0.82 J_\uparrow/k_B}$ and $\langle n_k^\downarrow \rangle_{E_0, dE}$ (same colouring scheme as in Fig. 2.9).

2.9 Comparison of the asymptotic state with thermal states

matches the crystal structure of the initial state. Hence the selection of excited states in the diagonal ensemble is fundamentally governed by the broken translational invariance present in the initial state. In contrast, the canonical ensemble averages over all eigenstates regardless of their displaying crystalline order, a fact which makes the loss of the crystalline structure unavoidable.

The density profiles (for the $\downarrow$-bosons) shown in Fig. 2.9 corroborate the previous statements. The diagonal ensemble induced by the trimer crystal is compared here with thermal averages in both the canonical and microcanonical ensembles. While the density profile in the diagonal ensemble still displays the "memory effect" of the initial crystalline state, the thermal states exhibit only small density modulations (in the microcanonical ensemble) or no modulation at all (in the canonical ensemble). Furthermore, the momentum profiles shown in Fig. 2.10 underline a non-thermalisation scheme of the time-evolved crystal state. While the density profile of the diagonal ensemble exhibits a pronounced peak at quasimomentum $k = 0$, this peak is almost completely washed out for the thermal ensembles.

In view of the two observables discussed here, a thermalisation of the evolved trimer crystal state can be excluded, at least for the finite-size system we are considering. This confirms the observations of "non-thermalisation" in other one-dimensional finite-size systems [22, 23].

Our exact diagonalization study is limited to a small cluster, and it cannot exclude a priori that thermalisation appears for larger system sizes: this would require that the diagonal ensemble converges to the microcanonical one, which ultimately converges to the canonical ensemble in the thermodynamic limit.

2.10 Numerical results of long-time evolutions

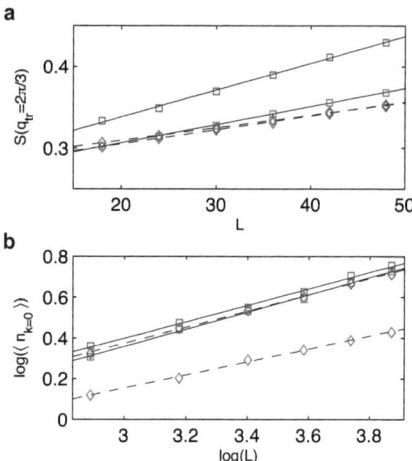

Figure 2.11: **Scaling analysis of the long-time evolution data.** (a) Structure factor peak $S(q_{tr} = 2\pi/3)$; (b) Quasi-condensate peak $\langle n_{k=0} \rangle$. Boxes (diamonds) stand for particle species $\downarrow$ ($\uparrow$), respectively. Parameters: $J_\downarrow/J_\uparrow = 0.15, U/J_\uparrow = 2.5$ (blue symbols) and $J_\downarrow/J_\uparrow = 0.40, U/J_\uparrow = 9.0$ (red symbols).

Here we present an example of the scaling analysis for the results of a long-time evolution up to $\tau = 150\hbar/J_\downarrow$. Fig. 2.11 shows that observables averaged over the last $\tau/4$ interval of the time evolution display the characteristic one-dimensional supersolid scaling, analogous to – but much more marked than – the one observed at short times (compare Fig. 2.5). Indeed we observe a linear scaling of the structure factor peak $S(q_{tr} = 2\pi/3)$ with system size, typical of solid order, and an algebraic sub-linear scaling of the condensed atoms, signaling quasi-condensation. Repeating this scaling analysis for a fine mesh of parameter space leads to the confirmation of the supersolid phase shown in Fig. 2.4.

A word of caution is necessary in the case of long-time evolutions. The truncation of the Hilbert space, inherent in all numerical algorithms for time evolution not based on full exact diagonalization [19], has the general effect that the accuracy of the results rapidly degrades with time, and the instantaneous measurements become practically unreliable in the long-time limit. We observe, however, that observables averaged over time intervals $> \tau/10$ do converge with high precision upon variation of the bond dimension D. These time averaged results are indeed the object of the above scaling analysis.

2.11 Conclusions

In summary, we have shown that supersolidity can appear dynamically in a bosonic mixture trapped in an optical lattice, showing commensurate crystalline order. This theoretical finding can be tested with presently available experimental techniques and may lead to the first observation of supersolidity.

References

[1] A. J. Leggett, Phys. Rev. Lett. **25**, 1543 (1970).

[2] G.V. Chester, Phys. Rev. A **2**, 256 (1970).

[3] E. Kim, M.H.W. Chan, Nature **427**, 225-227 (2004).

[4] E. Kim, M.H.W. Chan, Science **305**, 1941 (2004).

[5] J. Day, J. Beamish, Nature **450**, 853-856 (2007).

[6] N. Profok'ev, Adv. Phys. **56**, 381 (2007), and references therein.

[7] M.H.W. Chan, Science **319**, 120 (2008), and references therein.

[8] I. Bloch, J. Dalibard, W. Zwerger, Rev. Mod. Phys. **80**, 885 (2008).

[9] D. Jaksch, Nature **442**, 147-149 (2006), and references therein.

[10] I. Titvinidze, M. Snoek, W. Hofstetter, Phys. Rev. Lett. **100**, 100401 (2008).

[11] F. Hebert,G.G. Batrouni, X. Roy, Phys. Rev. B **78** 184505 (2008).

[12] D. Jaksch, C. Bruder, J.I. Cirac, C.W. Gardiner, P. Zoller, Phys. Rev. Lett. **81**, 3108 (1998).

[13] B. Paredes, *et al.*, Nature **429**, 277 (2004).

[14] O. Mandel *et al.*, Phys. Rev. Lett. **91**, 010407 (2003).

[15] L.M. Falicov, J.C. Kimball, Phys. Rev. Lett. **22**, 997 (1969).

[16] G.E. Astrakharchik, J. Boronat, J. Casulleras, S. Giorgini, Phys. Rev. Lett. **95**, 190407 (2005).

[17] S. Fölling et al., Nature **448**, 1029-1032 (2007).

[18] T. Giamarchi, *Quantum Physics in one dimension*. Clarendon Press, Oxford (2003).

[19] J.J. Garcia-Ripoll, New J. Phys. **8**, 305 (2006).

[20] G. Vidal, Phys. Rev. Lett. **93**, 040502 (2004).

[21] M. Rigol, V. Dunjko, M. Olshanii, Nature **452**, 854 (2008).

[22] C. Kollath, A. Läuchli, E. Altman, Phys. Rev. Lett. **98**, 180601 (2007).

[23] S.R. Manmana, S. Wessel, R.M. Noack, A. Muramatsu, Phys. Rev. Lett. **98**, 210405 (2007).

[24] A. Marte et al., Phys. Rev. Lett. **89**, 283202 (2002).

[25] C. Chin, R. Grimm, P. Julienne, *arXiv*:0812.1496 (2008), and references therein.

[26] J. Catani, et al., Phys. Rev. A **77**, 011603 (2008).

[27] G. Birkl, M. Gatzke, I.H. Deutsch, S.H. Rolston, W.D. Phillips, Phys. Rev. Lett. **75**, 2823 (1995).

[28] A.W. Sandvik, Phys. Rev. B **59**, R14157 (1999).

[29] T. Roscilde, Phys. Rev. A **77**, 063605 (2008).

Chapter 3

Dynamical creation of bosonic Cooper-like pairs

We propose a scheme to create a metastable state of paired bosonic atoms in an optical lattice. The most salient features of this state are that the wavefunction of each pair is a Bell state and that the pair size spans half the lattice, similar to fermionic Cooper pairs. This mesoscopic state can be created with a dynamical process that involves crossing a quantum phase transition and which is supported by the symmetries of the physical system. We characterize the final state by means of a measurable two-particle correlator that detects both the presence of the pairs and their size.

3.1 Introduction

Pairing is a central concept in many-body physics. It is based on the existence of quantum or classical correlations between pairs of components of a many-body system. The most relevant example of pairing is BCS superconductivity, in which attractive interactions cause electrons to perfectly anticorrelate in momentum and spin, forming Cooper pairs. In second quantization, this is described by the BCS

3. Dynamical creation of bosonic Cooper-like pairs

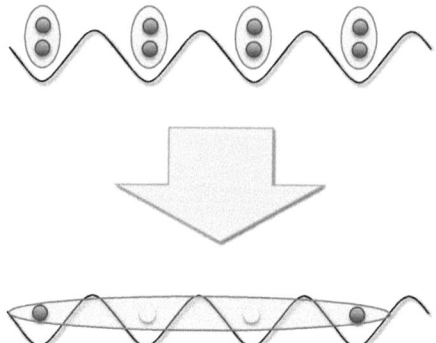

Figure 3.1: **Melting procedure of the entangled pair state.** The transition from the Mott into the superfluid regime does not need to be adiabatic, as the pairing is protected by entanglement.

variational wavefunction

$$|\psi_{\text{BCS}}\rangle = \prod_k (u_k + v_k A_k^\dagger)|0\rangle, \qquad (3.1)$$

where $A_k^\dagger \equiv c_{k\uparrow}^\dagger c_{-k\downarrow}^\dagger$ is an operator that creates one such Cooper pair. Remarkably, the fact that pairing occurs in momentum space means that the constituents of the pairs are delocalized and share some long-range correlation.

Nowadays, pairing and the creation of strongly correlated states of atoms is a key research topic. With the enhancement of atomic interactions due to Feshbach resonances, it has been possible both to produce Cooper pairs of fermionic atoms [1, 2, 3] and to observe the crossover from these large, delocalized objects to a condensate of bound molecular states. Realizing similar experiments with bosons is difficult, because attractive interactions may induce collapse. Two workarounds are based on optical lattices, either loaded with hard-core bosonic atoms [4] or, as in recent experiments [5], with metastable localized pairs supported by strong repulsive interactions.

In this work we propose a method to dynamically create long-range pairs of bosons which, instead of attractive interactions, uses entangled states as a resource. The

3.1 Introduction

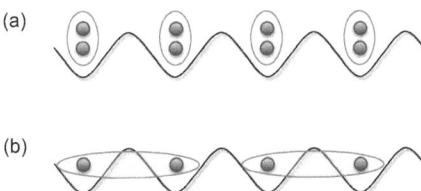

Figure 3.2: **Different initial states containing Bell pairs.**

method starts by loading an optical lattice of arbitrary geometry with entangled bosons that form an insulator. One possible family of initial states

$$|\psi\rangle \sim \prod_{i=1}^{L} A_{ii}^{\dagger}|0\rangle, \quad A_{ij} = \begin{cases} c_{i\uparrow}c_{j\uparrow} \pm c_{i\downarrow}c_{j\downarrow} \\ c_{i\uparrow}c_{j\downarrow} + c_{j\uparrow}c_{i\downarrow} \end{cases}, \quad (3.2)$$

are on-site pairs created by loading a lattice with two bosonic atoms per site and tuning their interactions, as demonstrated in Ref. [6]. A larger family includes states created by exchange interactions between atoms hosted in the unit cells of an optical superlattice [7, 8]

$$|\psi\rangle \sim \prod_{i=1}^{L/2} A_{2i-1,2i}^{\dagger}|0\rangle, \quad A_{ij} = \begin{cases} c_{i\uparrow}c_{j\uparrow} \pm c_{i\downarrow}c_{j\downarrow} \\ c_{i\uparrow}c_{j\downarrow} \pm c_{j\uparrow}c_{i\downarrow} \end{cases}. \quad (3.3)$$

We propose to dynamically increase the mobility of the atoms, entering the superfluid regime. During this process (see Fig. 3.1), pairs will enlarge until they form a stable gas of long-range Cooper-like pairs that span about half the lattice size. Contrary to works on the creation of squeezed states [9], the evolution considered here is not adiabatic and the survival of entanglement is ensured by a symmetry of the interactions.

This chapter is organized as follows. First, we present the Hamiltonian for bosonic atoms which are trapped in a deep optical lattice, have two degenerate internal states and spin independent interactions. Next, we prove that by lowering the optical lattice and moving into the superfluid regime, the Mott-Bell entangled states (3.2)-(3.3) evolve into a superfluid of pairs. We then introduce two correlators that detect the singlet and triplet pairs and their approximate size. These correlators are used to interpret quasi-exact numerical simulations of the evolution of two paired states as they enter the superfluid regime. Finally, we suggest two procedures to measure these correlators and elaborate on other experimental considerations.

3.2 Physical sytem

We will study an optical lattice that contains bosonic atoms in two different hyperfine states ($\sigma = \uparrow, \downarrow$). In the limit of strong confinement, the dynamics of the atoms is described by a Bose-Hubbard model [10]

$$H = -\sum_{\langle i,j \rangle, \sigma} J_\sigma c_{i\sigma}^\dagger c_{j\sigma} + \sum_{i\sigma\sigma'} \frac{1}{2} U_{\sigma\sigma'} c_{i\sigma}^\dagger c_{i\sigma'}^\dagger c_{i\sigma'} c_{i\sigma}. \tag{3.4}$$

Atoms move on a d-dimensional lattice ($d = 1, 2, 3$) jumping between neighboring sites with tunneling amplitude J_σ, and interacting on-site with strength $U_{\sigma\sigma'}$. The Bose-Hubbard model has two limiting regimes. If the interactions are weak, $U \ll J$, atoms can move freely through the lattice and form a superfluid. If interactions are strong and repulsive, $U \gg J$, the ground state is a Mott insulator with particles pinned on different lattice sites.

As mentioned in the introduction, we want to design a protocol that begins with an insulator of localized entangled states (3.2)-(3.3) and, by crossing the quantum phase transition, produces a gas of generalized Cooper pairs of bosons. In our proposal we restrict ourselves to symmetric interactions and hopping amplitudes

$$U \equiv U_{\uparrow\uparrow} = U_{\downarrow\downarrow} = U_{\uparrow\downarrow} \geq 0; \; J \equiv J_\uparrow = J_\downarrow \geq 0. \tag{3.5}$$

This symmetry makes the system robust so that, even though bosons do not stay in their ground state, they remain a coherent aggregate of pairs, unaffected by collisional dephasing. We will formulate this more precisely in the following section.

3.3 Conservation of pairing

Let us take an initial state of the form given by either Eq. (3.2) or (3.3). If we evolve this state under the Hamiltonian (3.4), with time-dependent but symmetric interaction $U_{\sigma\sigma'} = U(t)$ and hopping $J_\sigma = J(t)$, the resulting state will have a paired structure at all times

$$|\psi(t)\rangle = \sum_\pi c(t;\pi) A^\dagger_{\pi_1 \pi_2} \cdots A^\dagger_{\pi_{2L-1} \pi_{2L}} |0\rangle, \tag{3.6}$$

where $c(t;\pi)$ are complex coefficients and the sum over π denotes all possible permutations of the indices.

The proof of this result begins with the introduction of a set of operators

$$C_{ij} := \sum_\sigma c^\dagger_{i\sigma} c_{j\sigma} \tag{3.7}$$

which form a simple Lie algebra

$$[C_{ij}, C_{kl}] = C_{il}\delta_{jk} - C_{kj}\delta_{il}. \tag{3.8}$$

The evolution preserves the commutation relations and maps the group onto itself. This is evident if we rewrite the Hamiltonian

$$H = -J \sum_{\langle i,j \rangle} C_{ij} + \frac{U}{2} \sum_i (C_{ii})^2. \tag{3.9}$$

The evolution operator satisfies a Schrödinger equation

$$i\hbar \frac{d}{dt} V(t) = H(t) V(t), \tag{3.10}$$

with initial condition $V(0) = \mathbb{I}$. Since the Hamiltonian only contains C_{ij} operators we conclude that $V(t)$ is an analytic function of these generators. Let us focus on the evolution of state (3.3), given by

$$|\psi(t)\rangle = V(t) \prod_{i=1}^{L/2} A^\dagger_{2i-1,2i} |0\rangle. \tag{3.11}$$

We will use the commutation relations between the generators of the evolution and the pair operators

$$[A_{ij}, C_{kl}] = \delta_{ik} A_{lj} + \delta_{jk} A_{il}, \tag{3.12}$$

which are valid for any of the pairs in Eq. (3.3). Formally, it is possible to expand the unitary operator $V(t)$ in terms of the correlators C_{ij} and commute all these operators to the right of the A's, where we use $C_{ij}|0\rangle = 0$ and recover Eq. (3.6). A similar proof applies to the on-site pairs (3.2).

A particular case is the abrupt jump into the non-interacting regime, $U = 0$. Integrating this problem with initial conditions (3.2) and (3.3) the evolved state becomes

$$|\psi(t)\rangle = \prod_{x=1}^{N} \sum_{i,j} w(i-x, j-x, t) A_{ij}^{\dagger}|0\rangle. \qquad (3.13)$$

The wavepackets form an orthogonal set of states, initially localized $w(i,j,0) \propto \delta_{ij}$ or $w(i,j,0) \propto \delta_{ij+1}$ and approaching a Bessel function for large times[1]. We remark that though the pair wavefunctions (3.6) and (3.13) include valence bond states, they are more general because particles may overlap or form triplets.

3.4 Characterisation of the evolved pair wavefunction

In a general case, computing the many-body pair wavefunction, $c(t; \pi)$, is an open problem. Nevertheless we can prove that the final state *does not become the ground state of the superfluid regime*, no matter how slowly one changes the hopping and interaction. For the states in (3.3) this is evident from the lack of translational invariance. Let us thus focus on the state (3.2) generated by $A_{ii} = c_{i\uparrow} c_{i\downarrow}$, which has an equal number of spin-up and down particles $N_{\uparrow,\downarrow} = N/2$. The ground state of the same sector in the superfluid regime, $U = 0$, is a number squeezed, two-component condensate [9]

$$|\psi_{\text{NN}}\rangle \propto \tilde{c}_{0\uparrow}^{\dagger N/2} \tilde{c}_{0\downarrow}^{\dagger N/2}|0\rangle, \qquad (3.14)$$

with

$$\tilde{c}_{0\sigma} = \frac{1}{\sqrt{L}} \sum_{i=1}^{L} c_{i\sigma}. \qquad (3.15)$$

---
[1] The normalization of $w(i,j,t)$ depends on the choice of the operator A_{ij}.

3.5 Pair correlations

We can also write this ground state as an integral over condensates with atoms polarized along different directions

$$|\psi_{\text{NN}}\rangle \propto \int d\theta\, e^{-iN\theta/2}(\tilde{c}^\dagger_{0\uparrow} + e^{i\theta}\tilde{c}^\dagger_{0\downarrow})^N |0\rangle. \tag{3.16}$$

When this state is evolved backwards in time, into the $J = 0$ regime, each condensate transforms into an insulator with different polarization yielding

$$|\psi_{\text{NN}}\rangle \xrightarrow{\text{MI}} \sum_{\vec{n},\sum n_k = N/2} \prod_k (c^\dagger_{k\uparrow})^{n_k} (c^\dagger_{k\downarrow})^{2-n_k} |0\rangle. \tag{3.17}$$

Since this state is not generated by the $A_{ii} = c_{i\uparrow}c_{i\downarrow}$ operators, we conclude that this particular state (3.2), when evolved into the superfluid, leaves the ground state. Furthermore, since different pairs in Eq. (3.2) are related by global rotations, this statement applies to all of them. Indeed, numerical simulations indicate that the evolved versions of (3.2) and (3.3) are no longer eigenstates of (3.4).

3.5 Pair correlations

For the rest of this chapter we focus on two important states: the triplet pairs generated on the same site [6] and the singlet pairs generated on neighboring sites [7, 8],

$$|\psi_T\rangle = \prod_{i=1}^{L} \frac{1}{2}(c^{\dagger 2}_{i\uparrow} + c^{\dagger 2}_{i\downarrow})|0\rangle, \text{ and} \tag{3.18}$$

$$|\psi_S\rangle = \prod_{i=1}^{L/2} \frac{1}{\sqrt{2}}(c^\dagger_{2i-1\uparrow}c^\dagger_{2i\downarrow} - c^\dagger_{2i-1\downarrow}c^\dagger_{2i\uparrow})|0\rangle, \tag{3.19}$$

respectively. For a cartoon version of these states, see Fig. 3.2. Our goal is to study the evolution of these states as the mobility of the atoms is increased, suggesting experimental methods to detect and characterize the pair structure. The main tools in our analysis are the following two-particle connected correlators

$$\begin{aligned}G^T_{ij} &:= \left\langle c^\dagger_{i\uparrow} c^\dagger_{j\uparrow} c_{i\downarrow} c_{j\downarrow} \right\rangle - \left\langle c^\dagger_{i\uparrow} c_{i\downarrow} \right\rangle \left\langle c^\dagger_{j\uparrow} c_{j\downarrow} \right\rangle, \\ G^S_{ij} &:= \left\langle c^\dagger_{i\uparrow} c^\dagger_{j\downarrow} c_{j\uparrow} c_{i\downarrow} \right\rangle - \left\langle c^\dagger_{i\uparrow} c_{i\downarrow} \right\rangle \left\langle c^\dagger_{j\downarrow} c_{j\uparrow} \right\rangle,\end{aligned} \tag{3.20}$$

3. Dynamical creation of bosonic Cooper-like pairs

combined in two different averages

$$G_{\Delta \geq 0} = \frac{1}{L-\Delta} \sum_{i=1}^{L-\Delta} G_{i,i+\Delta}, \quad \bar{G} = \sum_{\Delta=0}^{L-1} G_\Delta \qquad (3.21)$$

and what we call the pair size

$$R \equiv \frac{\sum_\Delta |\Delta| \times |G_\Delta|}{\sum_\Delta |G_\Delta|}. \qquad (3.22)$$

A variant of the correlator G^T has been used as a pairing witness for fermions [11]. We expect these correlators to give information about the pair size and distribution also in the superfluid regime. This can be justified rigorously for an abrupt jump into the superfluid, in which the pair wavepackets remain orthogonal and G_Δ and R characterize the spread of the wavefunctions $w(i,j,t)$. First, note that the single-particle expectation values such as $\langle c_{i\downarrow}^\dagger c_{i\uparrow} \rangle$ are exactly zero since $N_\uparrow$ and $N_\downarrow$ are even for the triplet state ψ_T and balanced for the singlet state ψ_S. Second, the two-particle correlators only have nonzero contributions where the destruction and creation operators cancelled and subsequently created the same pair. Combining Eqs. (3.20) and (3.13) gives

$$\begin{aligned} G_{ij}^T &= \sum_x |w(i-x,j-x,t)|^2 4, \\ G_{ij}^S &= -\sum_x |w(j-x,i-x,t)|^2, \end{aligned} \qquad (3.23)$$

where we have used the symmetry of the wavefunction, $w(i,j,t) = w(j,i,t)$. Particularized to the initial states, the triplet ψ_T gives $G_{ij}^T = \delta_{ij}$, $G_\Delta^T = \delta_{\Delta 0}$, $\bar{G}^T = 1$, and $R^T = 0$, as expected from on-site pairs. The singlet pairs described by ψ_S, on the other hand, yield a non-zero G_{ij}^S only if i and j are the indices of the two ends of a singlet pair. Thus $G_\Delta^S = -\frac{1}{2}\delta_{\Delta 1}$, $\bar{G}^S = -\frac{1}{2}$, and $R^S = 1$.

3.6 Numerical details

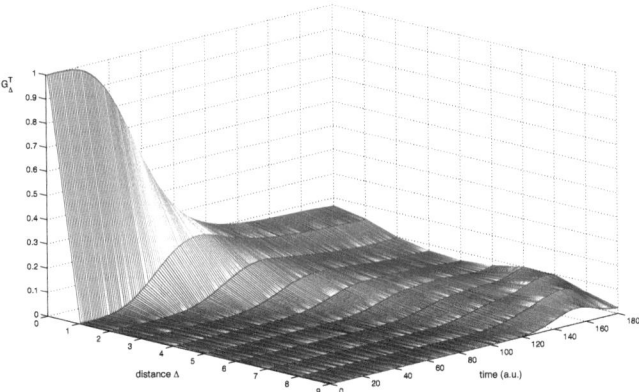

Figure 3.3: **Spread of the pair wave packet.** The correlator G_Δ^T is plotted as a function of the distance Δ and time t.

3.6 Numerical details

For a realistic study of the evolved paired states we have simulated the evolution of ψ_T and ψ_S under the Bose-Hubbard model as the hopping increases diabatically in time

$$J(t) = v \times (tU/\hbar) \times U, \qquad (3.24)$$

with ramp speeds $v = 0.5, 1$ and 2 in dimensionless units. The simulations were performed using Matrix Product States (MPS) on one-dimensional lattices with up to 20 sites and open boundary conditions [12, 13, 14]. After several convergence checks, we chose $D = 30$ for the MPS matrix size and $dt = 5 \times 10^{-4} \hbar/U$ for the time steps. For these small lattices, we expect the simulations to appropriately describe even the superfluid regime, where the small energy gaps and the high occupation number per site make the MPS algorithm more difficult.

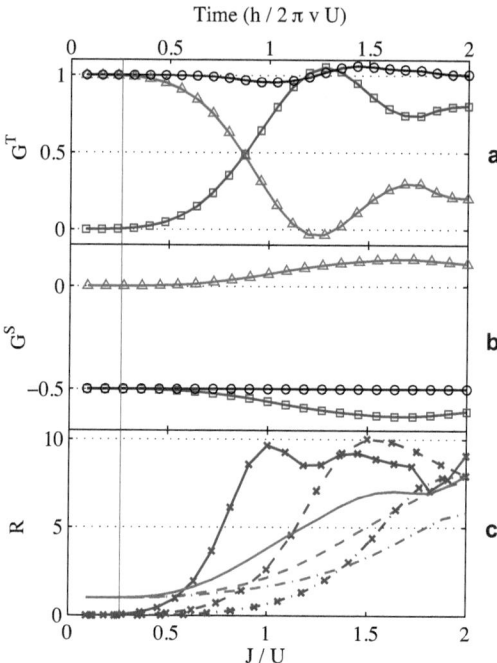

Figure 3.4: **Pair correlators and pair size as a function of evolution time.** We plot the (a) triplet and (b) singlet correlators for the evolution of ψ_T and ψ_S respectively, at a ramp speed $v = 1$ (see Eq. 3.24) and in a lattice of $L = 20$ sites. The circles, triangles and squares denote $\bar{G}$, G_0 and their difference. (c) Pair size R for the singlet (line) and triplet (cross) states, for a ramp speed $v = 0.5, 1$ and 2 (solid, dash, dash-dot). The vertical line $J/U = 1/3.84$ marks the location of the phase transition.

3.7 Cooper-like pairs of bosons

In Fig. 3.4 we plot the instantaneous values of the correlators and pair sizes along the ramp. Let us begin with the triplet pairs: initially the only relevant contribution is the short-range pair correlation G_0^T; then the pair size increases monotonously up to $R \sim L/2$, where it saturates. At this point, the pairs have become as large as the lattice permits, given that the density is uniform. The singlets have a slightly different dynamics. The antisymmetry of the spin wavefunction prevents two bosons of one pair to share the same site and thus $R = 1$ initially. This antisymmetry seems also to affect the overlap between pairs, as it is evidenced both in the slower growth $R(t)$ and in the smallness of G_0^S. Note that when the ramp is stopped (not shown here), the pair correlations persist but oscillate as the particles bounce off from each other and from the borders of the lattice.

Concerning the speed of the process, we have simulated ramps over a timescale which is comparable or even shorter than the typical interaction time, $1/U$, so that the process is definitely not adiabatic. Nevertheless, the pairs seem to have enough time to spread over these small lattices. Note also that the spreading of atoms begins right after the value $J/U \simeq 1/3.84$ where the one-dimensional Insulator-Superfluid phase transition takes place [15].

The system of delocalized Cooper-like pairs can also be regarded as a mean of distributing entanglement in the optical lattice. Following this line of thought we have used the von Neumann entropy to measure the entanglement between two halves of the optical lattice. A numerical study of the scaling of this entropy up to $L = 20$ sites, together with analytical estimates using the wavefunction (3.13) show that the entropy is far from the limit $O(N/2)$, which corresponds to perfectly splitting N distinguishable pairs among both lattice halves. We conjecture this is due to the pairs being composed of bosonic particles.

3.8 Measuring pair correlations

The pairing correlators $G^{T,S}$ can be decomposed into density-density correlations and measured using noise interferometry [16, 17]. To prove this, let us introduce the Schwinger representation of angular momenta

$$\begin{aligned}
S_x(i) &= \tfrac{1}{\sqrt{2}}(c_{i\uparrow}^\dagger c_{i\downarrow} + c_{i\downarrow}^\dagger c_{i\uparrow}), \\
S_y(i) &= \tfrac{1}{\sqrt{2}}(c_{i\uparrow}^\dagger c_{i\downarrow} - c_{i\downarrow}^\dagger c_{i\uparrow}), \\
S_z(i) &= \tfrac{1}{2}(c_{i\uparrow}^\dagger c_{i\uparrow} - c_{i\downarrow}^\dagger c_{i\downarrow}) = \tfrac{1}{2}(n_{i\uparrow} - n_{i\downarrow}).
\end{aligned} \quad (3.25)$$

For the states considered here, the correlation matrix is real (3.23). We can thus focus on its real part $\tilde{G}_{ij} = 2\mathrm{Re}(G_{ij})$, which is related to simple spin correlations

$$\begin{aligned}
\tilde{G}_{ij}^{T,S} &= \tfrac{1}{2}\langle S_x(i)S_x(j)\rangle \mp \tfrac{1}{2}\langle S_y(i)S_y(j)\rangle \\
&\quad - \tfrac{1}{2}\langle S_x(i)\rangle\langle S_x(j)\rangle \pm \tfrac{1}{2}\langle S_y(i)\rangle\langle S_y(j)\rangle.
\end{aligned} \quad (3.26)$$

We introduce two global rotations in the hyperfine space of the atoms

$$U_{x,y} = \exp\left[\pm i\frac{\pi}{2}\sum_k S_{y,x}(k)\right], \quad (3.27)$$

which take the S_x and S_y operators into the S_z, respectively. These rotations can be implemented experimentally without individual addressing and can be used to transform the spin correlators into density operators. For instance

$$\langle S_x(i)S_x(j)\rangle = \tfrac{1}{4}\left\langle U_x^\dagger (n_{i\uparrow}-n_{i\downarrow})(n_{j\uparrow}-n_{j\downarrow})U_x\right\rangle, \quad (3.28)$$

shows that the $S_x S_x$ arises from all possible density correlations after applying a $\pi/2$ pulse on the atoms.

Another possibility is to apply the ideas put forward in [18]. These methods rely on the interaction between coherent light and the trapped atoms to map quantum fluctuations of the atomic spin onto the light that crosses the lattice. Using this technique it should be possible to measure both the single-particle and the two-particle expectation values that constitute $G^{T,S}$.

Experimental imperfections are expected not to affect the nature of the final state. The influence of stray magnetic and electric fields can be obviated by working with the singlet pairs, which are insensitive to global rotations of the internal states and have large coherence times. More important could be the influence of any asymmetry in the interaction constants. However, assuming this asymmetry to be of the order of 1%, the effect can only be noticeable after a time $t = 100\hbar/U$, which is longer than the evolution times suggested here.

3.9 Conclusions

Summing up, in this chapter we have proposed a novel method to dynamically engineer Cooper pair like correlations between bosons. Our proposal represents a natural extension of current experiments with optical superlattices [7, 8]. It begins with a Mott insulator of bosonic atoms loaded in an optical lattice and forming entangled pairs, that have been created using quantum gates [6, 7, 8]. This state is diabatically melted into the superfluid regime so that the system becomes a stable gas of long-range correlated pairs. Unlike other systems, pairing is created dynamically, using entanglement as a resource, and supported by symmetries instead of attractive interactions. The generated states and the numerical and analytical tools developed in this work form a powerful toolbox to study, both experimentally and theoretically, issues like entanglement distribution and decoherence of many-body states. Future work will involve the quest for other resource states and diabatic protocols that lead to stronger correlations or more exotic states [19].

References

[1] C.A. Regal, M. Greiner, D.S. Jin, Phys. Rev. Lett. **92**, 040403 (2004).

[2] M.W. Zwierlein *et al.*, Phys. Rev. Lett. **92**, 120403 (2004).

[3] T. Bourdel *et al.*, Phys. Rev. Lett. **93**, 050401 (2004).

[4] B. Paredes, J.I. Cirac, Phys. Rev. Lett. **90**, 150402 (2003).

[5] K. Winkler *et al.*, Nature **441**, 853 (2006).

[6] A. Widera *et al.*, Phys. Rev. Lett. **92**, 160406 (2004).

[7] S. Fölling *et al.*, Nature **448**, 1029 (2007).

[8] M. Anderlini *et al.*, Nature **448**, 452 (2007).

[9] M. Rodriguez, S.R. Clark, D. Jaksch, Phys. Rev. A **75**, 011601 (2007).

[10] D. Jaksch *et al.*, Phys. Rev. Lett. **81**, 3108 (1998).

[11] C. Kraus, M.M. Wolf, J.I. Cirac, G. Giedke, Phys. Rev. A **79**, 012306 (2009).

[12] J.J. García-Ripoll, New J. Phys. **8**, 305 (2006).

[13] F. Verstraete, J.J. García-Ripoll, J.I. Cirac, Phys. Rev. Lett. **93**, 207204 (2004).

[14] G. Vidal, Phys. Rev. Lett. **93**, 040502 (2004).

[15] S. Rapsch, W. Zwerger, Europhys. Lett. **46**, 559 (1999).

[16] E. Altman, E. Demler, M.D. Lukin, Phys. Rev. A **70**, 013603 (2004).

[17] S. Fölling *et al.*, Nature **434**, 481 (2004).

[18] K. Eckert *et al.*, Nature Physics **4**, 50 (2008).

[19] B. Paredes, I. Bloch, Phys. Rev. A **77**, 023603 (2008).

Chapter 4

Pfaffian-like ground state for 3-body-hard-core bosons

We propose a Pfaffian-like Ansatz for the ground state of bosons subject to 3-body infinite repulsive interactions in a one-dimensional lattice. Our Ansatz consists of the symmetrization over all possible ways of distributing the particles in two identical Tonks-Girardeau gases. We support the quality of our Ansatz with numerical calculations and propose an experimental scheme based on mixtures of bosonic atoms and molecules in one-dimensional optical lattices in which this Pfaffian-like state could be realized. Our findings may open the way for the creation of non-Abelian anyons in one-dimensional systems.

4.1 Introduction

Beyond bosons and fermions, and even in contrast to the fascinating Abelian anyons (AA) [1], non-Abelian anyons (NAA) [2] exhibit an exotic statistical behavior: If two different exchanges are performed consecutively among identical NAA, the final state of the system will depend on the order in which the two exchanges were made. NAA appeared first in the context of the fractional quantum Hall effect (FQHE) [2], as elementary excitations of exotic states like the Pfaffian state [3, 4], the exact ground state of quantum Hall Hamiltonians with 3-body contact interactions. Recently, the possibility of a fault tolerant quantum computation based on NAA [5] has boosted the investigation of new models containing NAA [6], as well as the search for techniques for their detection and manipulation [7]. Meanwhile, the versatile and highly controllable atomic gases in optical lattices [3] have opened a door to the near future implementation of those models as well as for the artificial creation of non-Abelian gauge potentials [9].

All actual models containing NAA are 2D models. The motivation of the present work is the foreseen possibility of creating NAA in one dimension. Above all, this long-term goal requires the definition of the concept of NAA, which is essentially 2D, in one dimension. For Abelian anyons (AA) this generalization has already been made by Haldane [10]. Within his generalized definition the spinon excitations of one-dimensional Heisenberg antiferromagnets are classified as $\frac{1}{2}$-AA [10]. This classification becomes very natural through the connection between the one-dimensional antiferromagnetic ground state (for a long-range interaction model, the Haldane-Shastry model [11]) and the Laughlin state [5] for bosons at $\nu = 1/2$. In a similar way we anticipate that a connection can be established between quantum Hall models containing NAA and certain long-range one-dimensional spin models exhibiting NAA within a generalized definition.

Here, far from analyzing the above questions in general, our aim is to pave the way for the creation of exotic Pfaffian-like states in one-dimensional systems, which we believe may serve as the basis to create NAA. We present a realistic one-dimensional system whose ground state is very close to a Pfaffian-like state. The actual system we consider is that of bosonic atoms in a one-dimensional lattice with infinite repulsive 3-body on-site interactions, which we call *3-hard-core bosons*. Inspired by the form of the fractional quantum Hall Pfaffian state for bosons [4, 13], we propose an Ansatz for the ground state of our system. This Ansatz is a symmetrization over all possible ways of distributing the particles in two identical Tonks-Girardeau (T-G) gases [5, 15].

Comparison of the Ansatz with numerical calculations for lattices up to 40 sites yields very good agreement. As for fractional quantum Hall systems, NAA may be created here by creating pairs of quasiholes, each quasihole being in a different cluster [4]. This possibility will be discussed elsewhere.

Three-body collisions among single atoms rarely occur in nature. However, they can be effectively simulated by mixtures of bosonic particles and molecules. This has been proposed by Cooper [16] for a rapidly rotating gas of bosonic atoms and molecules. Here, we show that a system of atoms and molecules in a one-dimensional lattice can in a similar way effectively model 3-hard-core bosons. We will show that the conditions for realizing this situation lie within current experimental capabilities.

4.2 System setup

We consider a system of bosonic atoms in a one-dimensional lattice with repulsive 3-body on-site interactions. This system is described by the Hamiltonian:

$$H = -t \sum_{\ell} (a_{\ell}^{\dagger} a_{\ell+1} + h.c.) + U_3 \sum_{\ell} (a_{\ell}^{\dagger})^3 (a_{\ell})^3, \tag{4.1}$$

where the operator $a_{\ell}^{\dagger}$ (a_{ℓ}) creates (annihilates) a boson on site ℓ, t is the tunneling probability amplitude, and U_3 is the on-site interaction energy. From now on we will consider the limit $U_3 \to \infty$. In this limit the Hilbert space is projected onto the subspace of states with occupation numbers $n_{\ell} = 0, 1, 2$ per site. We will refer to bosons subject to this condition as 3-hard-core bosons. The projected Hamiltonian has the form

$$H_3 = -t \sum_{\ell} (a_{3,\ell}^{\dagger} a_{3,\ell+1} + h.c.), \tag{4.2}$$

where the 3-hard-core bosonic operators $a_{3,\ell}$ obey $(a_{3,\ell})^3 = 0$ and satisfy the commutation relations

$$[a_{3,\ell}, a_{3,\ell'}^{\dagger}] = \delta_{\ell,\ell'} \left(1 - \frac{3}{2}(a_{3,\ell}^{\dagger})^2 (a_{3,\ell})^2\right). \tag{4.3}$$

These operators can be represented by 3×3 matrices of the form

$$a_{3,\ell} = \begin{pmatrix} 0 & 1 & 0 \\ 0 & 0 & \sqrt{2} \\ 0 & 0 & 0 \end{pmatrix}. \tag{4.4}$$

In contrast to the usual hard-core bosonic operators [17], which are directly equivalent to spin-1/2 operators, the operators $a_3^\dagger$, a_3 are related to spin-1 operators $\{S^+, S^-, S^z\}$ in a non-linear way:

$$a_3 \to S^+ \left(\frac{1}{\sqrt{2}} + (\frac{1}{\sqrt{2}} - 1)S^z \right). \tag{4.5}$$

This mapping leads to a complicated equivalent spin Hamiltonian (with third and fourth order terms) which seems hard to solve.

4.3 Ansatz wavefunction for the ground state

In the following we present an Ansatz wave function for the ground state of Hamiltonian (4.2). Our Ansatz is inspired by the form of the ground state for fractional quantum Hall bosons subject to three body interactions [3, 4, 13]. The reason to believe that this inspiration may be good is the deep connection already demonstrated for the case of two-body interactions between ground states of certain one-dimensional models and those of 2D particles in the lowest Landau level (LLL) [11].

Let us now turn for a moment to the problem of bosons in the LLL subject to the 3-body interaction potential [3]

$$\sum_{i \neq j \neq k} \delta^2(z_i - z_j)\delta^2(z_i - z_k), \tag{4.6}$$

with $z_i = x_i + iy_i$ being the complex coordinate in the 2D plane. For infinite interaction strength the exact ground state of the problem is the Pfaffian state [3, 13]:

$$\Phi_3 \propto \mathcal{S}_{\uparrow,\downarrow} \left(\prod_{i<j}^{N/2}(z_i^\uparrow - z_j^\uparrow)^2 \prod_{i<j}^{N/2}(z_i^\downarrow - z_j^\downarrow)^2 \right). \tag{4.7}$$

4.3 Ansatz wavefunction for the ground state

This state is constructed in the following way. Particles are first arranged into two identical $\nu = 1/2$ Laughlin states [5],

$$\Phi_2^\sigma \propto \prod_{i<j}^{N/2} (z_i^\sigma - z_j^\sigma)^2 \qquad (4.8)$$

labeled by $\sigma = \uparrow, \downarrow$. Then the operator $\mathcal{S}_{\uparrow,\downarrow}$ symmetrizes over the two "virtual" subsets of coordinates $\{z_i^\uparrow\}$ and $\{z_i^\downarrow\}$. Note that the Laughlin state Φ_2^σ of each cluster is a zero energy eigenstate of the 2-body interaction potential $\sum_{i \neq j} \delta(z_i - z_j)$. This guarantees that in a state of the form (4.7) three particles can never coincide in the same position: for any trio, two of them will belong to the same group and cause the wave function Φ_3 to vanish.

In direct analogy with equation (4.7) we propose the following Ansatz for the ground state of Hamiltonian (4.2):

$$\Psi_3 \propto \mathcal{S}_{\uparrow,\downarrow} \left(\prod_{i<j}^{N/2} \left| \sin(x_i^\uparrow - x_j^\uparrow) \right| \prod_{i<j}^{N/2} \left| \sin(x_i^\downarrow - x_j^\downarrow) \right| \right). \qquad (4.9)$$

This Ansatz has the same structure as (4.7), but the Laughlin state has been substituted by a Tonks-Girardeau (T-G) state [5],

$$\Psi_2^\sigma \propto \prod_{i<j}^{N/2} \left| \sin(x_i^\sigma - x_j^\sigma) \right|, \qquad (4.10)$$

with $x_i^\sigma = 2\pi/Mi$, $i = 1, \ldots, M$; M being the number of lattice sites. This state is the ground state of hard-core one-dimensional lattice bosons with Hamiltonian

$$H_{2,\sigma} = -t \sum_\ell (a_{2,\sigma,\ell}^\dagger a_{2,\sigma,\ell+1} + h.c.) \qquad (4.11)$$

and periodic boundary conditions [15]. Here, $a_{2,\sigma}$ are hard-core bosonic operators satisfying $(a_{2,\sigma})^2 = 0$. Written in second quantization the Ansatz (4.9) takes the form:

$$|\Psi_3\rangle = \mathcal{P} \left(|\Psi_2^\uparrow\rangle \otimes |\Psi_2^\downarrow\rangle \right), \qquad (4.12)$$

where $\mathcal{P}$ is a local operator of the form $\mathcal{P} = \mathcal{P}_\ell^{\otimes M}$, and $\mathcal{P}_\ell$ is an operator mapping the single-site 4-dimensional Hilbert space of two species of hard-core bosons to the 3-dimensional one of 3-hard-core bosons (see Fig. 4.1).

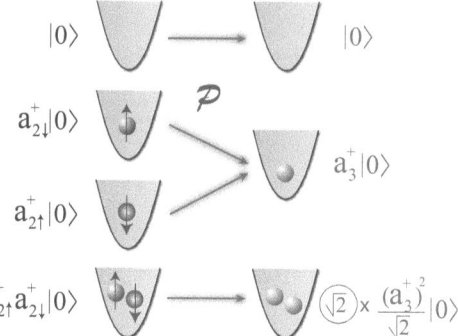

Figure 4.1: **Schematic representation of the operator** $\mathcal{P}_\ell$ mapping the single-site 4-dimensional Hilbert space of two species of hard-core bosons to the 3-dimensional one of 3-hard-core bosons.

4.4 Characterisation of the Ansatz wavefunction

Let us analyze different characteristic properties of our Ansatz. Taking into account the well known result for a T-G gas, namely the scaling of the one-particle correlation function

$$\langle a^\dagger_{\ell+\Delta} a_\ell \rangle \propto \Delta^{-1/2} \tag{4.13}$$

for large Δ [15], we can derive the following asymptotic behavior for the one-body and two-body correlation functions for the Ansatz (4.9):

$$\langle a^\dagger_{\ell+\Delta} a_\ell \rangle \longrightarrow \Delta^{-1/4} \tag{4.14}$$

$$\langle a^\dagger_{\ell+\Delta} a^\dagger_{\ell+\Delta} a_\ell a_\ell \rangle \longrightarrow \Delta^{-1}. \tag{4.15}$$

The result (4.15) can be derived by noticing that

$$\begin{aligned}
\langle a^\dagger_{\ell+\Delta} a^\dagger_{\ell+\Delta} a_\ell a_\ell \rangle &\propto \langle \Psi^\uparrow_2 | a^\dagger_{\ell+\Delta,\uparrow} a_{\ell,\uparrow} | \Psi^\uparrow_2 \rangle \langle \Psi^\downarrow_2 | a^\dagger_{\ell+\Delta,\downarrow} a_{\ell,\downarrow} | \Psi^\downarrow_2 \rangle \\
&\longrightarrow \Delta^{-1/2} \Delta^{-1/2}.
\end{aligned} \tag{4.16}$$

The proof of (4.14) is more involved and we will give just numerical evidence from our calculations below (see inset of Fig. 4.4). In our case, the two-body correlation (4.15) is indeed the (one-particle) correlation function for on-site pairs. This means that the system seen as a whole exhibits some kind of coherence (the spatial correlation decaying slowly as $\Delta^{-1/4}$), whereas the underlying system of on-site pairs is in a much more disordered state (with a fast correlation decay as Δ^{-1}). This is in contrast to what happens in a weakly interacting bosonic gas in which coherence between sites is independent of their occupation number. We can also obtain analytical expressions for the relative occupation of single and doubly occupied sites. The average number of doubly occupied sites is

$$n_2 = \langle a_\ell^\dagger a_\ell^\dagger a_\ell a_\ell \rangle / 2 = \nu^2/2, \qquad (4.17)$$

and the one of single occupied sites is given by

$$n_1 = \langle a_\ell^\dagger a_\ell (2 - n_\ell) \rangle = \nu(2 - \nu). \qquad (4.18)$$

This distribution is very different from the Poissonian one, for which we have $n_2^{\text{Po}}/n_1^{\text{Po}} = \nu/2$.

As an additional property, the Ansatz (4.9) has particle-hole symmetry. This means that for a filling factor of the form $\nu = N/M = 2 - \eta$, with N the number of particles, the state we propose is just the Ansatz for holes at $\nu_h = \eta$. However, as we can clearly see from its matrix representation, the Hamiltonian (4.2) does not exhibit this symmetry. This tells us that our Ansatz may not work for the whole regime of filling factors, as we will see in the next section.

4.5 Numerical details

In order to determine the quality of our Ansatz we have performed a numerical calculation for the ground state of Hamiltonian (4.2). To obtain the numerical ground state $|\Psi_{\text{ex}}\rangle$ we have used variational Matrix Product States (MPS) [18], with bond dimension $D = 15$, and physical dimension $d = 3$. We can estimate the error of this

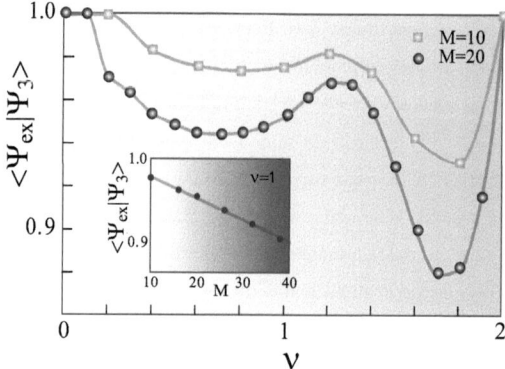

Figure 4.2: **Overlap** $\langle \Psi_{\text{ex}} | \Psi_3 \rangle$. The main plot shows the dependence of the overlap on the filling factor $\nu = N/M$ for a system of $M = 10$ (orange circles) and $M = 20$ (brown circles) lattice sites. The inset shows the decrease of the overlap with increasing system size M, at $\nu = 1$.

calculation to be smaller than 10^{-5} for the system sizes ($M \leq 40$) we have considered. To calculate the overlap of $|\Psi_{\text{ex}}\rangle$ with our Ansatz $|\Psi_3\rangle$ we first construct the MPS state that best approximates $|\Psi_3\rangle$ for a given D. This is done in the following way. We first build the MPS ground state of Hamiltonian H_2. We then take the tensor product of this state with itself, obtaining a MPS with $d = 4$, which is closest to $|\Psi_2^\uparrow\rangle \otimes |\Psi_2^\downarrow\rangle$. The dimension of the matrices of this state is very large and we use a reduction algorithm [19] to reduce it to the initial size. Finally we apply the operator $\mathcal{P}$ by local tensor contraction and normalize the resulting $d = 3$ MPS state. For the matrix dimensions we used the error made was always smaller than 10^{-3}.

4.6 Quality of the Ansatz

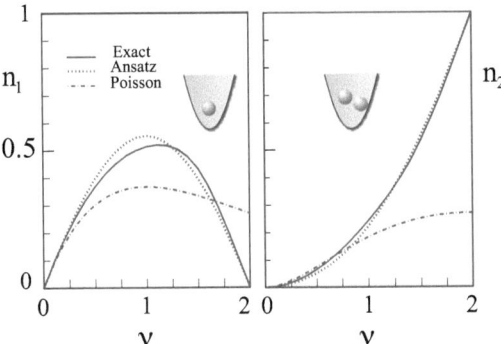

Figure 4.3: **Average occupation of sites** with one (left figure) and two particles (right figure) for the exact (solid lines), Ansatz (dotted lines) and Poissonian distribution (dot-dashed lines), as a function of the filling factor ν. The system size is $M = 20$.

4.6 Quality of the Ansatz

The results are shown in Fig. 4.2. The main plot shows the overlap between the Ansatz and the exact ground state $\langle \Psi_{\text{ex}} | \Psi_3 \rangle$ as a function of the filling factor $\nu = N/M$ for a fixed system size. We find very good overlaps (0.98-0.96) for $\nu \leq 1.25$. For $\nu > 1.25$ the overlap decreases. The inset shows the overlap as a function of increasing system size M, at fixed filling factor $\nu = 1$. At $M = 38$, the maximum size we have considered numerically, the overlap is still good (≈ 0.90).

Fig. 4.3 shows the statistical distribution of doubly and single occupied sites for $|\Psi_{\text{ex}}\rangle$, which is very close to the one of the Ansatz $|\Psi_3\rangle$, and clearly different from the Poissonian distribution typical of a weakly interacting Bose gas. Fig. 4.4 displays the momentum distribution for particles

$$n_k \propto \sum_{\ell,\Delta} e^{-ik\Delta} \langle a^\dagger_{\ell+\Delta} a_\ell \rangle \tag{4.19}$$

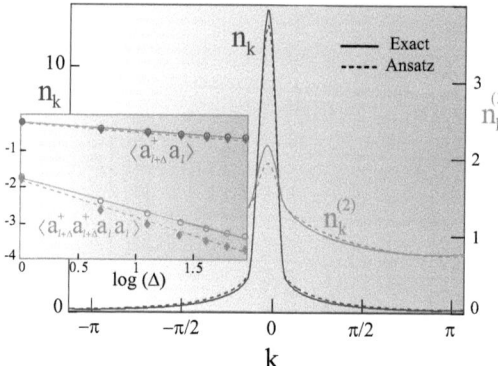

Figure 4.4: **Quasi-momentum distribution** of particles n_k (orange, left axis) and of on-site pairs $n_k^{(2)}$ (blue, right axis), with $k = 2\pi/Mn$, $n = 0, \ldots, M-1$. Results are shown both for the exact ground state (solid lines) and the Ansatz (dashed lines). The inset shows the long-distance scaling of the correlation functions $\langle a_{\ell+\Delta}^\dagger a_\ell \rangle \sim \Delta^{-\alpha_1}$ (orange), and $\langle a_{\ell+\Delta}^\dagger a_{\ell+\Delta}^\dagger a_\ell a_\ell \rangle \sim \Delta^{-\alpha_2}$ (blue), for the exact ground state (circles, $\alpha_1 = 0.22$, $\alpha_2 = 0.83$) and the Ansatz (diamonds, $\alpha_1 = 0.24$, $\alpha_2 = 0.99$. Parameters are $M = 20$ and $\nu = 1$.

and on-site pairs

$$n_k^{(2)} \propto \sum_{\ell,\Delta} e^{-ik\Delta} \langle a_{\ell+\Delta}^\dagger a_{\ell+\Delta}^\dagger a_\ell a_\ell \rangle, \qquad (4.20)$$

together with the long-range scaling of their spatial correlation functions. We can clearly see how the exact state exhibits all characteristic behaviours that we have discussed above for the Ansatz.

4.7 Experimental proposal

Inspired by Cooper's ideas [16] for 2D rotating Bose gases we present an experimental scheme for the realization of Hamiltonian (4.2). Let us consider a system of bosonic atoms and diatomic Feshbach molecules trapped in a one-dimensional optical lattice. The Hamiltonian of the system is $H = H_K + H_F + H_I$ [20, 21], where

$$\begin{aligned} H_K &= -t_a \sum_i (a_i^\dagger a_{i+1} + \text{h.c.}) - t_m \sum_i (m_i^\dagger m_{i+1} + \text{h.c.}), \\ H_F &= \sum_i \Delta m_i^\dagger m_i + \frac{U_{aa}}{2} a_i^\dagger a_i^\dagger a_i a_i + \frac{g}{\sqrt{2}} (m_i^\dagger a_i a_i + \text{h.c.}), \\ H_I &= U_{am} \sum_{i=1}^M m_i^\dagger a_i^\dagger a_i m_i + \frac{U_{mm}}{2} \sum_{i=1}^M m_i^\dagger m_i^\dagger m_i m_i. \end{aligned} \quad (4.21)$$

Here, the bosonic operators for atoms (molecules) a_i (m_i) obey the usual canonical commutation relations. The Hamiltonian H_K describes the tunneling processes of atoms and molecules, occurring with amplitude t_a and t_m, respectively. The term H_F is the Feshbach resonance term, with Δ being the energy off-set between open and closed channels, U_{aa} the on-site atom-atom interaction and g the coupling strength to the closed channel. Hamiltonian H_I describes the on-site atom-molecule and molecule-molecule interactions. We will assume a situation in which

$$U_{aa}, U_{am}, U_{mm} \geq 0, \quad (4.22)$$

and $\Delta > 0$. Furthermore, we will consider the limit in which

$$\gamma^2 = g^2/2\Delta^2 \ll 1. \quad (4.23)$$

Within this limit the formation of molecules is highly suppressed due to the high energy offset, Δ. However, virtual processes in which two atoms on the same lattice site go to the bound state, form a molecule and separate again, give rise to an effective

3-body interacting atomic Hamiltonian of the form[1]:

$$\begin{aligned}
H_{\text{eff}} = &- t_a \sum_i (a_i^\dagger a_{i+1} + \text{h.c.}) + U_{am}\gamma^2 \sum_i (a_i^\dagger)^3 (a_i)^3 \\
&- t_m \gamma^2 \sum_i \left((a_i^\dagger)^2 (a_{i+1})^2 + \text{h.c.} \right) \\
&+ (U_{aa} - g^2/\Delta) \sum_i (a_i^\dagger)^2 (a_i)^2,
\end{aligned} \qquad (4.24)$$

where we have neglected higher order terms in γ^2. Assuming $U_{aa} = g^2/\Delta$, and $t_m \gamma^2 \ll t_a$, H_{eff} reduces to Hamiltonian (4.1) with $t = t_a$ and $U_3 = U_{am}\gamma^2$. Finally, assuming $U_{am}\gamma^2 \gg t_a$ we end up with the desired Hamiltonian (4.2) for 3-hard-core bosons.

Let us now summarize the requirements and approximations we have imposed and discuss their experimental feasibility in typical setups with ^{87}Rb. We have assumed $g^2/\Delta = U_{aa}$. Since

$$g = \sqrt{U_{aa} \Delta\mu \Delta B / 2}, \qquad (4.25)$$

with ΔB being the width of the Feshbach resonance [20] and $\Delta\mu$ the difference in magnetic momenta, we need $\Delta = \Delta\mu \Delta B/2$. For the Feshbach resonance at $1007.4G$, this implies $\Delta/h = 441$kHz [22, 23].

Furthermore, we have assumed $\gamma^2 \ll 1$, $t_m \gamma^2 \ll t_a$ and $U_{am}\gamma^2/t_a \gg 1$. Written in terms of the lattice and atomic and molecule parameters we have

$$\begin{aligned}
\gamma^2 &= \sqrt{32/\pi^3} (a_{aa}^{3D} a / a_\perp^2)(\eta E_R / \Delta\mu \Delta B), \\
U_{am}/t_a &= (\sqrt{6}/4\pi)(a_{am}^{3D} a / a_\perp^2) \eta^{-2} \exp(+\pi^2 \eta^2/4), \\
t_m / t_a &= 2 \exp(-\pi^2 \eta^2/4),
\end{aligned} \qquad (4.26)$$

where $\eta = (V_0/E_R)^{1/4}$, with V_0 the lattice depth, $E_R = h^2/(8ma^2)$ the recoil energy, m the atomic mass, and a the lattice constant. The parameters a_{aa}^{3D} and a_{am}^{3D} are the 3D scattering lengths for atom-atom and atom-molecule collisions, and $a_\perp = \sqrt{\hbar/\omega_\perp m}$ is the transversal confinement width, with $\omega_\perp$ the transversal trapping frequency.

---

[1]This effective Hamiltonian is obtained by projection of Hamiltonian (4.21) onto the subspace with no molecules to first order in γ^2.

Assuming typical values $\eta^4 = 50(70)$, $a_{aa}^{3D} \sim 5nm$ [24], $a_{am}^{3D} \sim a_{aa}^{3D}$ [25], $a = 425nm$, and $\omega_\perp = 2\pi \times 20\text{kHz}$ [26], we obtain:

$$\gamma^2 = 3.55\,(3.86) \times 10^{-3},$$
$$U_{am}\gamma^2/t_a = 1.35\,(30.4) \times 10^3,$$
$$t_m/t_a = 5.29\,(0.22) \times 10^{-8}, \qquad (4.27)$$

clearly satisfying the required conditions.

Regarding detection of the Pfaffian-like ground state, the characteristic difference between both the momentum distribution and number statistics of particles and on-site pairs could be observed via spin-changing collisions [27].

4.8 Conclusions

In conclusion, we have shown that the ground state of 3-hard-core bosons in a one-dimensional lattice can be well described by a Pfaffian-like state which is a cluster of two T-G gases. We have shown that such a state may be accessible with current technology with atoms and molecules in optical lattices. We believe that our findings may open a new path for the creation of NAA.

References

[1] G.S. Canright and S.M. Girvin, Sciene **247**, 1197 (1990).

[2] G. Moore and N. Read, Nucl. Phys. B **360**, 362 (1991).

[3] M. Greiter, X.G. Wen, and F. Wilczek, Nucl. Phys. B **374**, 567 (1992);
C. Nayak and F. Wilczek, Nucl. Phys. B **479**, 529 (1996).

[4] N. Read, E. Rezayi, Phys. Rev. B **54**, 16864 (1996);
N. Read, E. Rezayi, Phys. Rev. B **59**, 8084 (1999).

[5] A.Y. Kitaev, Ann. Phys. (N.Y.) **303**, 2 (2003);
M. H. Freedman *et al.*, Commun. Math. Phys. **227**, 605 (2002).

[6] A.Y. Kitaev, Ann. Phys. **321**, 2 (2006).

[7] S. Das Sarma, M. Freedman, and C. Nayak, Phys. Rev. Lett. **94**, 166802 (2005);
A. Stern and B.I. Halperin, Phys. Rev. Lett. **96**, 016802 (2006);
P. Bonderson, A. Kitaev and K. Shtengel, Phys. Rev. Lett. **96**, 016803 (2006).

[8] I. Bloch, Physics World **17**, 25 (2004).

[9] K. Osterloh, M. Baig, L. Santos, P. Zoller, M. Lewenstein, Phys. Rev. Lett. **95**, 010403 (2005).

[10] F.D.M. Haldane, Phys. Rev. Lett. **67**, 937 (1991).

[11] F.D.M. Haldane, Phys. Rev. Lett. **60**, 635 (1988);
B.S. Shastry, Phys. Rev. Lett. **60**, 639 (1988).

[12] R.B. Laughlin, Phys. Rev. Lett. **50**, 1395 (1983).

[13] N.K. Wilkin and J.M.F. Gunn, Phys. Rev. Lett. **84**, 6 (2000).

[14] M. Girardeau, Journ. Math. Phys. **1**, 6 (1960).

[15] B. Paredes, A. Widera, V. Murg, O. Mandel, S. Fölling, I. Cirac, G.V. Shlyapnikov, T. W. Hänsch and I. Bloch, Nature **429**, 277 (2004).

[16] N.R. Cooper, Phys. Rev. Lett. **92**, 220405 (2004).

[17] S. Sachdev, *Quantum Phase Transitions* (Cambridge University Press, Cambridge, 1999).

[18] F. Verstraete, D. Porras, and J. I. Cirac, Phys. Rev. Lett. **93**, 227205 (2004).

[19] F. Verstraete, J.J. García-Ripoll, J.I. Cirac, Phys. Rev. Lett. **93**, 207204 (2004).

[20] E. Timmermans et al., Phys. Rev. Lett. **83**, 2691 (1999).

[21] M. Holland, J. Park, and R. Walser, Phys. Rev. Lett. **86**, 1915 (2001).

[22] T. Köhler, K. Goral, P.S. Julienne, Rev. Mod. Phys. **78**, 1311 (2006).

[23] S. Dürr, T. Volz, A. Marte, and G. Rempe, Phys. Rev. Lett. **92**, 020406 (2004).

[24] T. Volz, S. Dürr, S. Ernst, A. Marte, and G. Rempe, Phys. Rev. A **68**, 010702 (2003).

[25] S. Dürr, private communication.

[26] M. Greiner, I. Bloch, O. Mandel, T.W. Hänsch and T. Esslinger, Appl. Phys. B **73**, 769-772 (2001).

[27] A. Widera, F. Gerbier, S. Fölling, T. Gericke, O. Mandel, I. Bloch, Phys. Rev. Lett. **95**, 190405 (2005);
F. Gerbier, S. Fölling, A. Widera, O. Mandel, I. Bloch, Phys. Rev. Lett. **96**, 090401 (2006).

Chapter 5

Spin-charge separation in a one-dimensional spinor Bose gas

We study a one-dimensional (iso)spin 1/2 Bose gas with repulsive δ-function interaction by the Bethe Ansatz method and discuss the excitations above the polarized ground state. In addition to phonons the system features spin waves with a quadratic dispersion. We compute analytically and numerically the effective mass of the spin wave and show that the spin transport is greatly suppressed in the strong coupling regime, giving rise to a strong isospin-density (or "spin-charge") separation. Using a hydrodynamic approach, we study spin excitations in a harmonically trapped system and discuss prospects for future studies of two-component ultracold atomic gases.

5.1 Introduction

Recent experiments have shown the possibility of studying ultra-cold atomic gases confined in very elongated traps [1, 2, 3, 4]. In such geometries, the gas behaves kinematically as if it were truly one-dimensional. Many theoretical studies [5, 6, 7, 8, 9, 10] have predicted and discussed interesting effects in one-dimensional Bose gases, such as the occurrence of fermionization in the strong coupling Tonks-Girardeau (TG) regime, where elementary excitations are expected to be similar to those of a non-interacting one-dimensional Fermi gas [5]. Manifestations of strong interactions have been found in the experiments [2], and recently the TG regime has been achieved for bosons in an optical lattice [3] and in the gas phase [4].

Present facilities allow one to create spinor Bose gases which have been demonstrated in experimental studies of two-component Bose-Einstein condensates [11]. These systems are produced by simultaneously trapping atoms in two internal states, which can be referred to as (iso)spin 1/2 states. Relative spatial oscillations of the two components can be viewed as spin waves [12] (see [13] for a review). A variety of interesting spin-related effects such as phase separation [14], exotic ground states [15], and counter intuitive spin dynamics [12] due to the exchange mean field, have been studied both theoretically and experimentally. However, most of these studies are restricted to the weakly interacting Gross-Pitaevskii (GP) regime. There is a fundamental question to what extent these effects survive in the strongly correlated regime characteristic of one spatial dimension. The purpose of the present work is to study spin excitations of an interacting one-dimensional spinor Bose gas. This is done by employing an exact solution by the Bethe Ansatz.

5.2 System setup

We start with a spinor (two component) gas of N bosons with mass m at zero temperature, interacting with each other via a repulsive short-range potential in a narrow three-dimensional waveguide. In general, the interaction depends on the internal (spin) states of the colliding particles. Here we consider the case of a spin-independent interaction characterized by a single 3D scattering length $a > 0$. This is a reasonable approximation for the commonly used internal levels of ^{87}Rb (see e.g. [11]). The waveguide has length L and we assume periodic boundary conditions for simplicity. The transverse confinement is due to a harmonic trapping potential of frequency ω_0. When the chemical potential of the gas is much smaller than $\hbar\omega_0$, the transverse

5.2 System setup

motion is frozen to zero point oscillations with amplitude $l_0 = \sqrt{\hbar/m\omega_0}$. In such a quasi-one-dimensional geometry, the interaction between atoms is characterized by an effective one-dimensional delta-potential $g\delta(x)$. For $a \ll l_0$, the coupling constant g is related to the 3D scattering length as $g = 2\hbar^2 a/m l_0^2 > 0$ [8]. The behavior of the system depends crucially on the dimensionless parameter

$$\gamma = mg/\hbar^2 n, \qquad (5.1)$$

where $n = N/L$ is the one-dimensional density. For $\gamma \ll 1$ one obtains the weak coupling GP regime, whereas for $\gamma \gg 1$ the gas enters the strongly interacting TG regime.

Under the above conditions, the system is governed by the following spin-independent one-dimensional total Hamiltonian:

$$H = -\frac{\hbar^2}{2m}\sum_{i=1}^{N}\frac{\partial^2}{\partial x_i^2} + g\sum_{i<j}\delta(x_i - x_j). \qquad (5.2)$$

This Hamiltonian was introduced by Lieb and Liniger [6] for describing spinless bosons, and their solution by the Bethe Ansatz (BA) has been generalized to bosons or fermions in two internal states by M. Gaudin and C.N. Yang [16, 17]. In the case of a two-component Bose gas (spin 1/2 bosons), due to the $SU(2)$ symmetry of the Hamiltonian the eigenstates are classified according to their total (iso)spin S ranging from 0 to $N/2$. In this case, which was recently considered by Li, Gu, Yang and Eckern [18], the ground state is fully polarized ($S = N/2$) and has $2S+1$-fold degeneracy, in agreement with a general theorem [19, 20]. At a fixed $S = N/2$, the system is described by the Lieb-Liniger (LL) model [6], for which elementary excitations have been studied in [7] for any value of the interaction constant. Spin excitations above the ground state are independent of the ground-state spin projection M_S and represent transverse spin waves. For $M_S = 0$ they correspond to relative oscillations of the two gas components.

80   5. Spin-charge separation in a one-dimensional spinor Bose gas

5.3 Bethe Ansatz solution

We first give a brief summary of the BA diagonalization [16, 17, 18] of the Hamiltonian (5.2). An eigenstate with total spin

$$S = N/2 - K \ (0 \leq K \leq N/2) \tag{5.3}$$

is characterized by two sets of quantum numbers: N density quantum numbers I_j with $j = 1, .., N$ and K spin quantum numbers J_μ with $\mu = 1, .., K$.
If $N - K$ is odd (resp. even), I_j and J_μ are integers (resp. half-integers). These quantum numbers define N quasi-momenta k_j and K spin rapidities λ_μ, which satisfy the following set of BA equations (we set $\hbar = 2m = 1$):

$$\frac{Lk_j}{2} = \pi I_j - \sum_{l=1}^{N} \arctan\left(\frac{k_j - k_l}{g/2}\right) + \sum_{\nu=1}^{K} \arctan\left(\frac{k_j - \lambda_\nu}{g/4}\right), \tag{5.4}$$

$$\pi J_\mu = \sum_{l=1}^{N} \arctan\left(\frac{\lambda_\mu - k_l}{g/4}\right) - \sum_{\nu=1}^{K} \arctan\left(\frac{\lambda_\mu - \lambda_\nu}{g/2}\right). \tag{5.5}$$

The energy of the corresponding state is $E = \sum_j k_j^2$, and its momentum is given by:

$$p = \sum_{j=1}^{N} k_j = \frac{2\pi}{L}\Big(\sum_{j=1}^{N} I_j - \sum_{\mu=1}^{K} J_\mu\Big). \tag{5.6}$$

As we are also interested in finite size effects, we do not take the thermodynamic limit at this point.

The ground state corresponds to the quantum numbers

$$\{I_j^0\} = \{-(N-1)/2, .., (N-1)/2\} \tag{5.7}$$

and $K = 0$, which shows that the BA equations reduce to those of LL [6]. The wave function is given by the orbital wave function of the LL ground state multiplied by a fully polarized spin wave function. All ground state orbital properties (energy, chemical potential, correlation functions, etc.) are therefore identical to those of the LL model. Elementary excitations in the density sector correspond to modifying the density quantum numbers I_j while leaving the total spin unchanged, i.e., $K = 0$.

5.3 Bethe Ansatz solution

At low energy, the density excitations are phonons propagating with the Bogoliubov sound velocity $v_s = \sqrt{2gn}$ in the GP limit and with the Fermi velocity $v_s = 2\pi n$ in the TG regime.

We now focus on the spin sector. Elementary spin excitations correspond to reversing one spin ($K = 1$), and the total spin changes from $N/2$ to $N/2 - 1$. Thus, we have a single spin rapidity λ and the corresponding quantum number J. In general, this procedure creates a density excitation and a spin wave (isospinon) [18]. Here, we choose specific quantum numbers I_j, J in order to excite the isospinon alone[1]. Accordingly, the momentum p of the excitation is

$$p = \frac{2\pi}{L}\left(\frac{N}{2} - J\right), \quad (5.8)$$

which follows from the definition (5.6).

In the long wavelength limit, where $|p| \ll n$, due to the $SU(2)$ symmetry one expects [21] a quadratic dispersion for the spin-wave excitations above the ferromagnetic ground state:

$$\varepsilon_p \equiv E(p) - E_0 \simeq p^2/2m^*, \quad (5.9)$$

where $E(p)$ is the energy of the system in the presence of a spin wave with momentum p, E_0 is the ground state energy and m^* is an effective mass (or inverse spin stiffness). This quadratic behavior is due to a vanishing inverse spin susceptibility, which is a consequence of the $SU(2)$ symmetry [21]. A variational calculation in the spirit of Feynman's single mode approximation [20], shows that

$$\varepsilon_p \leq p^2/2m \quad (5.10)$$

implying $m^* \geq m$. Below we show that strong interactions greatly enhance the effective mass.

---

[1] In order to excite an isospinon alone we choose density quantum numbers $I_j = j - N/2$ and the spin quantum number $J = 1 - N/2 + l$, where l varies between 1 and $N - 2$ [18].

5.4 Strong coupling regime

In the strong coupling limit it is possible to solve the BA equations (5.4) and (5.5) perturbatively in $1/\gamma$ [6]. We solve these equations both for the ground state $\{I_j^0\}$ and the excited state $\{I_j; J\}$. We anticipate that in the limit of strong interactions, for small momenta ($|p|/n \ll 1$) and a large number of particles ($N \gg 1$), the dimensionless spin rapidity is

$$\tilde{\lambda} \equiv 2\lambda/g \gg 1 \qquad (5.11)$$

and the dimensionless quasi-momenta are $|k_j|/g \ll 1$. This allows us to expand Eqs. (5.4) and (5.5) to first order in $1/\gamma$ and $1/N$. The ground state quasi-momenta are then given by:

$$k_j^0 L = 2\pi I_j^0 \left(1 - 2/\gamma\right). \qquad (5.12)$$

Here we used the relation

$$\sum_l \arctan(2(k_j^0 - k_l^0)/g) \simeq 2Nk_j^0/g - 2(\sum_l k_l^0)/g = 2Nk_j^0/g, \qquad (5.13)$$

which is a consequence of the vanishing ground state momentum. Similarly, the excited state quasi-momenta obey the equations:

$$k_j L = \left(1 - \frac{2}{\gamma}\right) 2\pi I_j + \frac{2pL}{N\gamma} - \pi + \frac{1}{\tilde{\lambda}}\left(1 + \frac{k_j L}{\gamma N \tilde{\lambda}}\right), \qquad (5.14)$$

where p is given by Eq. (5.6). Neglecting quasi-momenta k_l in the argument of arctangent in the BA equation (5.5), we obtain the excited state spin rapidity:

$$2\pi J = 2N \arctan(2\tilde{\lambda}) \simeq \pi N - N/\tilde{\lambda} \qquad (5.15)$$

Equations (5.8) and (5.15) then give:

$$\tilde{\lambda} = N/pL, \qquad (5.16)$$

which justifies that $\tilde{\lambda} \gg 1$ for $|p|/n \ll 1$. Combining this result with Eq. (5.14) shows that $|k_j|/g \ll 1$, as anticipated. Let us now define the shift of the quasi-momenta $\Delta k_j \equiv k_j - k_j^0$. Taking the difference between equations (5.14) and (5.12), we find:

$$\Delta k_j = \frac{1}{L\tilde{\lambda}} + \frac{k_j^0}{\gamma N \tilde{\lambda}^2} + \frac{2p}{\gamma N} - \frac{2\pi}{L\gamma} \qquad (5.17)$$

where we used that $I_j - I_j^0 = 1/2$. We can now compute the energy of the spin wave, as defined in Eq. (5.9):

$$\varepsilon_p = \sum_{j=1}^{N} \left[2k_j^0 \Delta k_j + (\Delta k_j)^2 \right]. \tag{5.18}$$

Using Eq. (5.17) for Δk_j and Eq. (5.16) for $\tilde{\lambda}$ gives

$$\varepsilon_p = p^2 \left(1/N + 2\pi^2/3\gamma \right). \tag{5.19}$$

Note that the last two terms in the right hand side of Eq. (5.17) give no contribution, as the ground state momentum is zero. According to the definition (5.9), the inverse effective mass is therefore:

$$\frac{m}{m^*} = \frac{1}{N} + \frac{2\pi^2}{3\gamma}, \tag{5.20}$$

where we restored the units. Remarkably, the effective mass reaches the total mass Nm for $\gamma \to \infty$: the bosons are impenetrable and therefore a flipped spin boson can move on a ring only if all other bosons move as well.

5.5 Weak coupling regime

In the opposite limit of weak interactions it is possible to compute the effective mass from the Bogoliubov approach [22]. The validity of this procedure when considering a one-dimensional Bose gas, i.e. in the absence of a true Bose-Einstein condensate, is justified in [23]. The Hamiltonian of the system can be written as $H_0 + H_{int}$, where H_0 is the Hamiltonian of free Bogoliubov quasiparticles and free spin waves:

$$H_0 = \sum_p \epsilon_p \alpha_p^\dagger \alpha_p + \sum_p e_p \beta_p^\dagger \beta_p, \tag{5.21}$$

with α_p, β_p being the Bogoliubov quasiparticle and the spin wave field operators, $\epsilon_p = \sqrt{e_p(e_p + 2gn)}$ the Bogoliubov spectrum, and $e_p = p^2/2m$ the spectrum of free spin waves[2]. The Hamiltonian H_{int} describes the interaction between Bogoliubov

---
[2] The mechanism responsible for spin waves in the GP regime is the so-called "quantum torque" (see e.g. [13]).

quasiparticles and spin waves and provides corrections to the dispersion relations ϵ_p and e_p. The most important part of H_{int} reads:

$$H_{int} = g\sqrt{\frac{n}{L}} \sum_{k,q\neq 0} \left(u_q \alpha_q^\dagger - v_q \alpha_{-q}\right) \beta_{k-q}^\dagger \beta_k + \text{h.c.}, \quad (5.22)$$

where u_q and v_q are the u,v Bogoliubov coefficients satisfying the relations [22]

$$u_q + v_q = \sqrt{\epsilon_q/e_q} \quad (5.23)$$

and

$$u_q - v_q = \sqrt{e_q/\epsilon_q}. \quad (5.24)$$

Neglected terms contribute only to higher orders in the coupling constant. To second order in perturbation theory, in the thermodynamic limit the presence of a spin wave changes the energy of the system by:

$$\Delta E(p) = e_p + \frac{g^2 n}{2\pi\hbar} \int dq \, \frac{e_q}{\epsilon_q} \frac{1}{e_p - [\epsilon_q + e_{p+q}]}. \quad (5.25)$$

In order to calculate a correction to the effective mass of the spin wave, we expand Eq. (5.25) in the limit of $p \to 0$. Terms which do not depend on p modify the ground state energy, linear terms vanish, and quadratic terms modify the spin wave spectrum as follows:

$$\varepsilon_p = e_p \left(1 - \frac{4g^2 n}{\pi\hbar} \int_0^\infty dq \, \frac{e_q}{\epsilon_q} \frac{e_q}{[\epsilon_q + e_q]^3}\right), \quad (5.26)$$

where the main contribution to the integral comes from momenta $q \sim \sqrt{mgn}$. Using the definition (5.9), we then obtain the inverse effective mass:

$$\frac{m}{m^*} = 1 - \frac{2\sqrt{\gamma}}{\pi} \int_0^\infty dx \, \frac{(\sqrt{1+x^2} - x)^3}{\sqrt{1+x^2}} = 1 - \frac{2\sqrt{\gamma}}{3\pi}, \quad (5.27)$$

which clearly shows non-analytical corrections to the bare mass due to correlations between particles. This result can also be obtained directly from the BA equations.

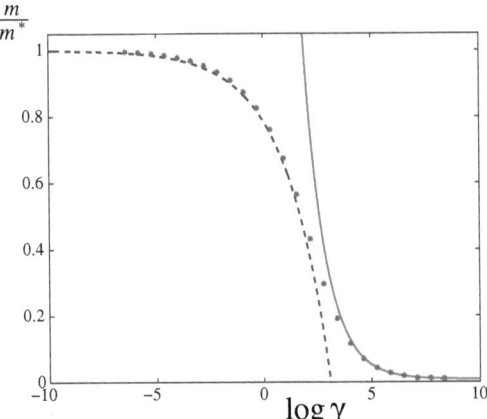

Figure 5.1: **Inverse effective mass m/m^* as a function of the dimensionless coupling constant γ (logarithmic scale).** The stars ($*$) show numerical results for $N = 111$ particles, the solid curve represents the behavior in the strong coupling limit (Eq. (5.20)), and the dashed curve the behavior for a weak coupling (Eq. (5.27)).

5.6 Numerical confirmation

For intermediate couplings, we obtained the effective mass by numerically solving the BA equations (5.4) and (5.5). Our results are displayed in Fig. 5.1. Note that when solving the BA equations, one should take care of choosing γ as

$$N^{-2} \ll \gamma \ll N^2. \tag{5.28}$$

Indeed, if $\gamma < N^{-2}$, the potential energy per particle in the weak coupling limit is lower than the zero point kinetic energy $\hbar^2/mL^2$ and the gas is therefore non-interacting (effectively $\gamma = 0$). In the strong coupling limit and for the same reason, if $\gamma > N^2$, the system behaves as a TG gas (effectively $\gamma = \infty$).

5.7 Hydrodynamical approach

We now turn to harmonically trapped bosons in the TG regime and rely on spin hydrodynamics introduced for uniform systems [21]. As the ground state is fully polarized we assume the equilibrium (longitudinal) spin density

$$\vec{S}(x) = n(x)\hat{e}_3 \tag{5.29}$$

and study small transverse spin density fluctuations

$$\delta\vec{S}(x,t) = \delta S_1 \hat{e}_1 + \delta S_2 \hat{e}_2, \tag{5.30}$$

where $\hat{e}_1, \hat{e}_2, \hat{e}_3$ form an orthonormal basis in the spin space. For a large N, the equilibrium density profile $n(x)$ in a harmonic trapping potential $V(x) = m\omega^2 x^2/2$ is given by the Thomas-Fermi expression

$$n(x) = n_0\sqrt{1 - (x/R)^2}. \tag{5.31}$$

Here $n_0 = n(0)$ is the density in the center of the trap and $R = \sqrt{2\hbar N/m\omega}$ is the Thomas-Fermi radius. For a strong but finite coupling Eq. (5.31) represents the leading term, with corrections proportional to inverse powers of $\gamma_0 = mg/\hbar^2 n_0$. The spin density fluctuations $\delta\vec{S}$ obey the following linearized Landau-Lifshitz equations [21]:

$$\delta\dot{S}_{1,2} = \mp\frac{\hbar}{2}\partial_x \frac{n(x)}{m^*(x)}\partial_x \frac{\delta S_{2,1}}{n(x)}. \tag{5.32}$$

In the TG regime the effective mass entering the equation of motion (5.32) depends on the density profile $n(x)$ as

$$m^*(x)/m \approx 3\gamma(x)/2\pi^2 = 3mg/2\pi^2\hbar^2 n(x). \tag{5.33}$$

Using the density profile (5.31) and introducing a complex function

$$n(x)\Phi(x,t) = \delta S_1(x,t) + i\delta S_2(x,t), \tag{5.34}$$

one obtains from Eqs. (5.32):

$$i\dot{\Phi} = \Omega\Phi = -\frac{\pi^2}{6}\frac{\omega}{\gamma_0 N}\frac{1}{\sqrt{1-X^2}}\partial_X\left(1-X^2\right)\partial_X\Phi, \tag{5.35}$$

where $X = x/R$ is the dimensionless coordinate, and we assumed the stationary time dependance

$$\Phi(X, t) = e^{-i\Omega t}\Phi(X). \quad (5.36)$$

Equation (5.35) shows that the typical frequency scale of the isospin excitations is given by $\omega/\gamma_0 N$, which is smaller than the scale ω of acoustic frequencies by a large factor $\gamma_0 N$. The exact solution to this equation was obtained numerically using the shooting method, and the spectrum shows only a small difference from the semi-classical result

$$\Omega_j = \frac{A\omega}{\gamma_0 N}\left(j + \frac{1}{2}\right)^2, \quad j = 0, 1, 2, \ldots, \quad (5.37)$$

where the numerical factor is $A = \pi^5/48\Gamma^4(3/4) \approx 2.83$. For $\omega \sim 100$ Hz, $\gamma_0 \sim 10$ and $N \sim 100$ as in the experiment [4], the lowest eigenfrequencies Ω_j are two or three orders of magnitude smaller than acoustic frequencies and are ~ 0.1 Hz.

5.8 Conclusions

In conclusion, we have found extremely slow (iso)spin dynamics in the strong coupling TG regime, originating from a very large effective mass of spin waves. In an experiment with ultra-cold bosons, this should show up as a spectacular isospin-density separation: an initial wave packet splits into a fast acoustic wave traveling at the Fermi velocity and an extremely slow spin wave [24]. One can even think of "freezing" the spin transport, which in experiments with two-component one-dimensional Bose gases will correspond to freezing relative oscillations of the two components.

References

[1] F. Schreck *et al.*, Phys. Rev. Lett. **87**, 080403 (2001);
A. Görlitz *et al.*, Phys. Rev. Lett. **87**, 130402 (2001);
M. Greiner *et al.*, Phys. Rev. Lett. **87**, 160405 (2001).

[2] H. Moritz *et al.*, Phys. Rev. Lett. **91**, 250402 (2003);
B. Laburthe Tolra *et al.*, Phys. Rev. Lett. **92**, 190401 (2004).

[3] B. Paredes *et al.*, Nature **429**, 277 (2004).

[4] T. Kinoshita, T. Wenger and D.S. Weiss, Science **305**, 1125 (2004).

[5] M. Girardeau, J. Math. Phys. **1**, 516 (1960).

[6] E.H. Lieb and W. Liniger, Phys. Rev. **130**, 1605 (1963).

[7] E.H. Lieb, Phys. Rev. **130**, 1616 (1963).

[8] M. Olshanii, Phys. Rev. Lett. **81**, 938 (1998).

[9] D.S. Petrov, G.V. Shlyapnikov and J.T.M. Walraven, Phys. Rev. Lett. **85**, 3745 (2000).

[10] V. Dunjko, V. Lorent and M. Olshanii, Phys. Rev. Lett. **86**, 5413 (2001).

[11] D.S. Hall *et al.*, Phys. Rev. Lett. **81**, 1539 (1998);
D.S. Hall *et al.*, Phys. Rev. Lett. **81**, 1543 (1998).

[12] H.J. Lewandowski *et al.*, Phys. Rev. Lett. **88**, 070403 (2002);
M.Ö. Oktel and L.S. Levitov, Phys. Rev. Lett. **88**, 230403 (2002);

J.N. Fuchs, D.M. Gangardt and F. Laloë, Phys. Rev. Lett. **88**, 230404 (2002);

J.E. Williams, T. Nikuni and C.W. Clark, Phys. Rev. Lett. **88**, 230405 (2002);

J.M. McGuirk et al., Phys. Rev. Lett. **89**, 090402 (2002).

[13] T. Nikuni and J.E. Williams, J. Low Temp. Phys. **133**, 323 (2003).

[14] W.B. Colson and A. Fetter, J. Low Temp. Phys. **33**, 231 (1978);

T.-L. Ho and V.B. Shenoy, Phys. Rev. Lett. **77**, 3276 (1996);

B.D. Esry et al., Phys. Rev. Lett. **78**, 3594 (1997);

C.K. Law et al., Phys. Rev. Lett. **79**, 3105 (1997);

H. Pu and N.P. Bigelow, Phys. Rev. Lett. **80**, 1130 (1998).

[15] T.-L. Ho and S.K. Yip, Phys. Rev. Lett. **84**, 4031 (2000);

A.B. Kuklov and B.V. Svistunov, Phys. Rev. Lett. **89**, 170403 (2002).

[16] M. Gaudin, Phys. Lett. A **24**, 55 (1967).

[17] C.N. Yang, Phys. Rev. Lett. **19**, 1312 (1967).

[18] Y.Q. Li et al., Europhys. Lett. **61**, 368-374 (2003).

[19] E. Eisenberg and E.H. Lieb, Phys. Rev. Lett. **89**, 220403 (2002).

[20] K. Yang and Y.Q. Li, Int. J. Mod. Phys. B **17**, 1027 (2003).

[21] B.I. Halperin and P.C. Hohenberg, Phys. Rev. **188**, 898 (1969);

B.I. Halperin, Phys. Rev. B **11**, 178 (1975).

[22] See, e.g., E.M. Lifshitz and L.P. Pitaevskii, *Statistical Physics, Part 2* (Butterworth-Heinemann, 1980).

[23] V.N. Popov, *Functional Integrals in Quantum Field Theory and Statistical Physics* (Reidel, Dordrecht, 1983).

[24] We find that in the one-dimensional spinor Bose gas the "spin-charge separation" is significantly stronger than in the other proposal for spin $1/2$ atomic *fermions*, see A. Recati et al., Phys. Rev. Lett. **90**, 020401 (2003).

Chapter 6

Anyons in one-dimensional optical lattices

In the following we present an exact mapping of Abelian anyons in a one-dimensional optical lattice onto bosons with conditional hopping amplitudes. An experimental implementation of this specific bosonic Hamiltonian would realize a gas of anyons. We will outline a possible experimental setup using laser-assisted tunneling. This work is still in preparation and has not been published yet.

6.1 Introduction

Ordinarily, every particle in quantum theory is neatly classified as either a boson – a particle happy to fraternize with any number of identical particles in a single quantum state – or a fermion, which insists on sole occupancy of its state. In the space of three or more dimensions, particles are restricted to being fermions or bosons, according to their statistical behaviour. While fermions obey Fermi-Dirac statistics, bosons are required to respect Bose-Einstein statistics. The interchange of two fermions leads – due to the Pauli principle – to a phase factor -1 in the total wavefunction, while the wavefunction of two bosons remains invariant under particle exchange.

6. Anyons in one-dimensional optical lattices

Almost 30 years ago, researchers proposed a third fundamental category of particles living in two-dimensional systems, "anyons" [1, 2, 3]. For two anyons, the wavefunction acquires a fractional phase $e^{i\theta}$ under particle interchange, giving rise to "fractional statistics", with $0 < \theta < \pi$. Note that the excluded cases $\theta = 0$ ($\theta = \pi$) would yield bosonic (fermionic) statistics. **Any**ons can indeed be **any**where inbetween those two limiting cases, giving rise to infinitely many fractional phases and thus an extraordinary wealth of exotic quantum particles.

At present, anyons or fractional statistics have not been detected directly in experiments. However, a significant fraction of physicists believe that the many-particle state of electrons observed in the so-called fractional quantum Hall (FQH) effect can potentially qualify as an anyonic playground. In the FQH effect [4, 5], each electron traveling in a thin conducting strip pierced by a strong magnetic field becomes associated with several vortices, tiny whirlpools of electric current flowing around field lines. Several experiments demonstrated that the vortices can be identified as quasiparticles each having a fraction of the electron's charge, but at present the fractional statistics of those quasiparticles could not be verified.

For a few years the habitat for anyons remained restricted to the two-dimensional world [6], until Haldane presented the concept of fractional statistics in arbitrary dimensions [7]. In this article, fractional statistics is reformulated as a generalization of the Pauli exclusion principle, and a definition independent of the dimension of space is obtained. Indeed, the a priori definition of anyonic statistics (see next section) does not imply any dimensional restrictions.

In the context of ultracold atomic setups for creating anyons, we briefly wish to review two promising theoretical proposals. Cold atoms in rapidly rotating fields have been predicted to exhibit fractional statistics [8], when the rotation frequency reaches a certain threshold. The physical models describing either charged particles in strong magnetic fields (related to the FQH) or neutral particles in rapidly rotating fields are mathematically equivalent, i.e. the Lorentz and Coriolis forces have the same vectorial structure. The authors in [8] show how anyonic excitations can be created and moved

6.1 Introduction

around by piercing the rotating atomic cloud with lasers. Unfortunately, the rapid rotation frequencies required for this proposal to work have not been experimentally achieved so far.

In another, more recent work [9], the authors discuss the possibility of creating fractional statistics in a one-dimensional Hubbard-model of fermions with "bond-charge" interaction. This interaction can be regarded as tunneling of the fermionic $\uparrow$-species, conditional on the presence of a $\downarrow$-fermion on the source or target sites. This model realizes fractional elementary excitations on a one-dimensional lattice, which can be identified as anyons.

In the following, we will establish an exact mapping between anyons and bosons in one dimension, via a generalized Jordan-Wigner transformation. We will show that anyons moving on a one-dimensional lattice are equivalent to – and can be realized by – ordinary bosons with conditional hopping amplitudes. This work is still unfinished. At this stage, we will present the analytical heart of this project, which proves the mapping between anyons and "conditional-hopping bosons" on a lattice. Furthermore, we will present an outlook concerning the realization of this specific bosonic model, discussing an experimental scheme involving laser-assisted, state-dependent tunneling, by which the fractional phase θ can be directly controlled.

6.2 Anyon statistics

Let us introduce the anyonic commutation relations

$$a_i a_j^\dagger - f(\theta) a_j^\dagger a_i = \delta_{ij}, \qquad (6.1)$$

with the phase factor carrying the fractional statistics

$$f(\theta) = e^{i\theta \operatorname{sgn}(i-j)}. \qquad (6.2)$$

The operators $a_i^\dagger, a_i$ create, annihilate an anyon on site i respectively. Note that the sign function is defined as follows

$$\operatorname{sgn}(i-j) = \begin{cases} +1, & i > j \\ 0, & i = j \\ -1, & i < j \end{cases} \qquad (6.3)$$

For the case $i = j$ the anyonic commutation relation eq. 6.1 thus reduces to bosonic ones $a_i a_i^\dagger - a_i^\dagger a_i = 1$.

6.3 Fractional Jordan-Wigner mapping

In the following we introduce an exact mapping between anyons and bosons, using a generalized Jordan-Wigner transformation.

Let us define

$$a_i = b_i e^{i\theta \sum_{k<i} n_k} \qquad (6.4)$$

with $n_k = a_k^\dagger a_k = b_k^\dagger b_k$ the number operator for both particle types. Provided that the particles of type b are bosons, $[b_i, b_j^\dagger] = \delta_{ij}$, we will prove in the following that the mapped operators a indeed obey the anyonic commutation relations from eq. 6.1.

For the case $i < j$ we wish to rewrite products of anyonic operators in terms of the bosonic ones:

$$\begin{aligned}
a_i a_j^\dagger &= b_i e^{-i\theta \sum_{i \leq k < j} n_k} b_j^\dagger \\
&= e^{-i\theta \sum_{i<k<j} n_k} b_i b_j^\dagger e^{-i\theta n_i}, \\
f(\theta) a_j^\dagger a_i &= e^{-i\theta \sum_{i<k<j} n_k} e^{-i\theta n_i} f(\theta) b_j^\dagger b_i \\
&= e^{-i\theta \sum_{i<k<j} n_k} e^{-i\theta(n_i+1)} b_j^\dagger b_i.
\end{aligned} \quad (6.5)$$

Here we have used $f(\theta) = e^{-i\theta}$ since $i < j$ was assumed. We can now evaluate the LHS of eq. 6.1:

$$\begin{aligned}
a_i a_j^\dagger - f(\theta) a_j^\dagger a_i &= e^{-i\theta \sum_{i<k<j} n_k} (b_i b_j^\dagger e^{-i\theta n_i} - e^{-i\theta(n_i+1)} b_j^\dagger b_i) \\
&= e^{-i\theta \sum_{i<k<j} n_k} e^{-i\theta(n_i+1)} [b_i, b_j^\dagger] \\
&= 0
\end{aligned} \quad (6.6)$$

Thus the anyonic commutation relations have been proven for the case $i < j$. The proof for the case $i > j$ is very similar. For the case $i = j$ one just has to note that $a_i^\dagger a_i = b_i^\dagger b_i$ and $f(\theta) = 1$.

6.4 Anyons mapped onto bosons

Our ultimate goal is to propose a realizable setup for demonstrating a gas of anyons in one dimension. The Hamiltonian that describes this system is

$$H^a = -J \sum_i (a_i^\dagger a_{i+1} + \text{h.c.}) + U \sum_i n_i, \quad (6.7)$$

with J the tunneling and U the on-site interaction amplitudes for anyons. By using our anyon-boson mapping eq. (6.4), we can rewrite the anyonic Hamiltonian H^a simply in terms of bosonic operators:

$$H^b = -J \sum_i (b_i^\dagger b_{i+1} e^{i\theta n_i} + \text{h.c.}) + U \sum_i n_i. \quad (6.8)$$

The mapped, bosonic Hamiltonian thus describes bosons with a conditional hopping amplitude $J e^{i\theta n_i}$. If the target site is unoccupied, the hopping amplitude is simply J. If it is occupied by one boson, the amplitude reads $J e^{i\theta}$, and so on.

6.5 Restoring left-right symmetry

Note that the left-right symmetry is broken in eq. (6.8), due to the phase factor acting only on the left site i. This symmetry is broken by the choice of the "tail orientation" in the Jordan-Wigner transformation, eq. (6.4). Thus by working with two versions of transformations, one with a tail in the left and one in the right direction, one can restore the symmetry for the final Hamiltonian. We thus define

$$\begin{aligned} a_{i,L} &= b_i e^{i\theta \sum_{k<i} n_k}, \\ a_{i,R} &= b_i e^{i\theta \sum_{k>i} n_k}. \end{aligned} \quad (6.9)$$

Using these two definitions we arrive at two bosonic Hamiltonians

$$\begin{aligned} H_L^b &= -J \sum_i (b_i^\dagger b_{i+1} e^{i\theta n_i} + \text{h.c.}) + U \sum_i n_i, \\ H_R^b &= -J \sum_i (b_{i+1}^\dagger b_i e^{i\theta n_{i+1}} + \text{h.c.}) + U \sum_i n_i. \end{aligned} \quad (6.10)$$

Both versions are simply related by an exchange $i \leftrightarrow i+1$. We can symmetrize over left and right versions to finally obtain

$$\begin{aligned} H_S^b &= 1/2(H_L^b + H_R^b) \\ &= -J/2 \sum_i (b_i^\dagger b_{i+1} e^{i\theta n_i} + b_{i+1}^\dagger b_i e^{i\theta n_{i+1}} + \text{h.c.}) \\ &\quad + U \sum_i n_i, \end{aligned} \quad (6.11)$$

which is left-right symmetric.

6.6 Experimental proposal

Let us imagine a situation, in which the lattice site occupations are restricted to $n_i = 0, 1, 2$. This situation can be realized by a strong on-site interaction or a dilute gas of bosons trapped in the optical lattice potential.

Let us now focus on the Hamiltonian with broken left-right symmetry (parity), eq. (6.8). In the truncated Hilbert space we only have to distinguish two cases:

6.6 Experimental proposal

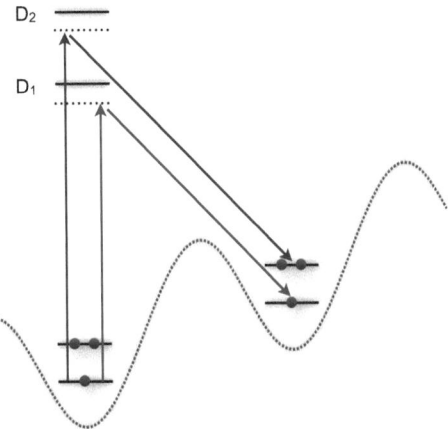

Figure 6.1: **Laser-assisted, state-dependent tunneling scheme.**

If the target site is unoccupied, the tunneling amplitude is J. If it is already occupied by one boson, the amplitude reads $Je^{i\theta}$. (If the target site is occupied by two bosons, no tunneling is possible.) Which experimental setup could potentially be sensitive to those two distinct cases, and at the same time give rise to broken parity? A laser-assisted, conditional tunneling scheme, similar to the one proposed by Jaksch in 2003 [10], seems highly promising.

Fig. 6.1 shows the basic idea. The optical lattice is tilted, this additional homogeneous field breaks parity. Ground states with different occupations and internally excited states (taken to be the D_1 and D_2 states of ^{87}Rb) are shown. The "red" Raman process selects the transition from a singly occupied source to a singly occupied target site, via the excited D_1 level, thus giving rise to the tunneling transition

$$|1\rangle_i \to |1\rangle_{i+1}, \qquad (6.12)$$

where $i, i+1$ label the source and target sites.

Similarly, the "blue" Raman process yields the other transition we need (via the D_2 level),

$$|1\rangle_i \to |2\rangle_{i+1}. \qquad (6.13)$$

In this way, one can drive the two different tunneling transitions by adjusting the frequencies of the external lasers. Furthermore, by phase locking the "blue" with respect to the "red" lasers, the phase θ (and thus the anyon statistics) can be easily tuned. In this sense, the tunneling dependent on the target site's occupation can be implemented in present optical lattice setups. This laser-assisted tunneling scheme is therefore a good candidate to realize the Hamiltonian (6.8), and thus Abelian anyons.

Note that the processes shown in Fig. 6.1 only couple to the source site with single occupation. The extension of this scheme to both singly and doubly occupied source sites is straight-forward, one just needs to add two pairs of lasers with the correct frequencies.

This scheme can be in principle generalized also to higher occupation numbers $n_i > 2$, one just needs to add more lasers driving all the different tunneling processes.

6.7 Conclusions

In this chapter we have presented a way to realize Abelian anyons. We have shown that Abelian anyons in one-dimensional lattices are equivalent to bosons, that feature a hopping amplitude conditional on the target site's occupation. We have proposed a way to realize this specific Hamiltonian using Raman transitions driven by external laser fields, allowing the anyon statistics angle to be tuned freely. This work is still unpublished.

References

[1] J. Leinaas and J. Myrheim, Nuovo Cimento B **37**, 1 (1977).

[2] G.A. Goldin, R. Menikoff, and D.H. Sharp, J. Math. Phys. **22**, 1664 (1981).

[3] F. Wilczek, Phys. Rev. Lett. **48**, 1144 (1982).

[4] D.C. Tsui, H.L. Stormer and A.C. Gossard, Phys. Rev. Lett. **48**, 1559 (1982).

[5] R.B. Laughlin, Phys. Rev. Lett. **50**, 1395 (1983).

[6] G.S. Canright and S.M. Girvin, Sciene **247**, 1197 (1990).

[7] F.D.M. Haldane, Phys. Rev. Lett. **67**, 937 (1991).

[8] B. Paredes, P. Fedichev, J. I. Cirac and P. Zoller, Phys. Rev. Lett. **87**, 010402 (2001).

[9] C. Vitoriano and M. D. Coutinho-Filho, Phys. Rev. Lett. **102**, 146404 (2009).

[10] D. Jaksch and P. Zoller, New J. Phys. **5**, 56 (2003).

REFERENCES

List of Figures

2.1 **Schematic of the quantum quench leading to supersolidity.** A product state of bosonic trimers is the initial state of the evolution (larger symbols represent the $\downarrow$-bosons); switching off one of the superlattice components leads to a supersolid state in which the particles delocalize into a (quasi-)condensate while maintaining the original solid pattern without imperfections. 22

2.2 **Dynamical onset of supersolidity by quantum quenching a mixture of light and heavy bosons.** Momentum profile of the $\downarrow$-bosons, $\langle n_k^\downarrow \rangle$ vs. time in units of hopping events $\hbar/J_\downarrow$. A quasi-condensate peak develops rapidly. Inset: Density distribution $\langle n_i^\downarrow \rangle$ averaged over the last third of the evolution time, showing that crystalline order is conserved in the system. The simulation parameters are $L = 28$, $N_\downarrow = 18$, $N_\uparrow = 9$, $J_\downarrow/J_\uparrow = 0.1, U/J_\uparrow = 3.0$. 23

2.3 **Equilibrium phase diagram (ground state).** The dash-dotted line represents the points where the hopping of the $\downarrow$-bosons, $J_\downarrow$, overcomes the energy gap to crystal dislocations, giving rise to the solid/super-Tonks (s-Tonks) transition. The dashed line marks the points where a single-trimer wavefunction spreads over 2.8 sites. In the super-Tonks phase, quasi-solidity and superfluidity coexist. 25

2.4 **Out-of-equilibrium phase diagram.** An extended supersolid phase exists in the transient state attained after the quantum quench. In this phase true solidity and quasi-condensation coexist. Blue symbols delimit the boundaries of the solid phase, red symbols mark the lower boundary for the quasi-condensed (q-c) phase. The overlap of both phases (blue shaded region) is identified as the supersolid phase. The yellow-filled symbols correspond to equilibrium data points. The lower boundary of the superfluid/super-Tonks region of the equilibrium phase diagram is seen to coincide with the lower boundary of the supersolid region out of equilibrium. 28

2.5 **Coexistence of solid order and quasi-condensation in the supersolid phase.** (a) The structure factor peak $S(q_{tr} = 2\pi/3)$ scales linearly with system size L, demonstrating solid order for both bosonic species. (b) The density peak in momentum space $\langle n_{k=0}^{\downarrow} \rangle$ is plotted vs. L on a log-log scale, showing algebraic scaling and thus quasi-condensation. Boxes (diamonds) stand for particle species $\downarrow$ ($\uparrow$), respectively. The data represented by blue boxes in part (a) is offset by -0.2 for better visibility. Parameters: $J_\downarrow/J_\uparrow = 0.1, U/J_\uparrow = 3.0$ (blue symbols) and $J_\downarrow/J_\uparrow = 0.15, U/J_\uparrow = 2.5$ (red symbols). 29

2.6 **Snapshot of a supersolid.** Square modulus of the natural orbital $\chi_i^{(0)}$ corresponding to the largest eigenvalue of the OBDM, calculated at final time τ. In the supersolid regime (blue/red symbols for $\downarrow/\uparrow$ bosons), the natural orbital shows the characteristic crystalline order. This pattern is washed out in the purely quasi-condensed regime (dashed/solid curves for $\downarrow/\uparrow$). The supersolid data is offset by +0.02 for the sake of visibility. Parameters: $J_\downarrow/J_\uparrow = 0.1$ (supersolid), $J_\downarrow/J_\uparrow = 0.8$ (quasi-condensed), $U/J_\uparrow = 3.0$, $N_\downarrow = 18$, $N_\uparrow = 9$, $L = 28$. 30

LIST OF FIGURES

2.7 **Overlap of the equilibrium ground state with the initial trimer-crystal state.** The overlap $|c_0|^2$ (contour plot) agrees well with the boundaries of the non-equilibrium supersolid phase (black symbols, cf. Fig. 2.4). This suggests a superfluid ground state as a necessary condition for supersolidity to dynamically set in. The overlap $|c_0|^2$ has been calculated via exact diagonalization on a $L = 10$ chain containing three trimers. 31

2.8 **Diagonal vs. thermal probability distributions.** The occupations of the diagonal ($|c_a|^2$ in blue) and canonical ($|d_a|^2$ in green) ensembles are plotted as a function of the eigenstate energies (offset from E_{GS}). Contrary to the thermal, continuous distribution, the trimer-crystal state emphasizes certain eigenstates, while it suppresses others. The (superfluid) ground state contribution present in the trimer-crystal state is enhanced by a factor of ≈ 20 compared with the thermal contribution. Most of the amplified excited states indeed show a crystalline structure with the correct periodicity, or contain density peaks at the right positions to build up the final crystal. Inset: The same distributions on a log-lin scale. The deviation of the diagonal from the thermal ensemble is even better visualized here. 36

2.9 **Real-space density $\langle n_i^\downarrow \rangle$ in all three ensembles.** While the diagonal ensemble $\langle n_i^\downarrow \rangle_\infty$ (blue) shows a clear crystalline pattern, this structure is washed out completely in the canonical ensemble $\langle n_i^\downarrow \rangle_{T=0.82 J_\uparrow / k_B}$ (green). Results for the microcanonical ensemble $\langle n_i^\downarrow \rangle_{E_{\text{in}}, dE}$ are shown for energy windows $dE = 0.2 J_\uparrow$ (red) and $dE = 0.6 J_\uparrow$ (cyan). All thermal ensembles deviate strongly from the density structure at time $t \to \infty$ (diagonal ensemble). 38

2.10 **Momentum profile $\langle n_k^\downarrow \rangle$ in all three ensembles.** Due to the significant weight attributed to the ground state, the diagonal ensemble $\langle n_k^\downarrow \rangle_\infty$ (blue) features an enhanced quasi-condensation peak at $k = 0$. This peak is suppressed in all thermal ensembles $\langle n_k^\downarrow \rangle_{T=0.82J_\uparrow/k_B}$ and $\langle n_k^\downarrow \rangle_{E_0,dE}$ (same colouring scheme as in Fig. 2.9). 38

2.11 **Scaling analysis of the long-time evolution data.** (a) Structure factor peak $S(q_{tr} = 2\pi/3)$; (b) Quasi-condensate peak $\langle n_{k=0} \rangle$. Boxes (diamonds) stand for particle species $\downarrow$ ($\uparrow$), respectively. Parameters: $J_\downarrow/J_\uparrow = 0.15, U/J_\uparrow = 2.5$ (blue symbols) and $J_\downarrow/J_\uparrow = 0.40, U/J_\uparrow = 9.0$ (red symbols). 40

3.1 **Melting procedure of the entangled pair state.** The transition from the Mott into the superfluid regime does not need to be adiabatic, as the pairing is protected by entanglement. 46

3.2 **Different initial states containing Bell pairs.** 47

3.3 **Spread of the pair wave packet.** The correlator G_Δ^T is plotted as a function of the distance Δ and time t. 53

3.4 **Pair correlators and pair size as a function of evolution time.** We plot the (a) triplet and (b) singlet correlators for the evolution of ψ_T and ψ_S respectively, at a ramp speed $v = 1$ (see Eq. 3.24) and in a lattice of $L = 20$ sites. The circles, triangles and squares denote $\bar{G}$, G_0 and their difference. (c) Pair size R for the singlet (line) and triplet (cross) states, for a ramp speed $v = 0.5, 1$ and 2 (solid, dash, dash-dot). The vertical line $J/U = 1/3.84$ marks the location of the phase transition. 54

4.1 **Schematic representation of the operator $\mathcal{P}_\ell$** mapping the single-site 4-dimensional Hilbert space of two species of hard-core bosons to the 3-dimensional one of 3-hard-core bosons. 66

LIST OF FIGURES

4.2 **Overlap** $\langle\Psi_{\text{ex}}|\Psi_3\rangle$. The main plot shows the dependence of the overlap on the filling factor $\nu = N/M$ for a system of $M = 10$ (orange circles) and $M = 20$ (brown circles) lattice sites. The inset shows the decrease of the overlap with increasing system size M, at $\nu = 1$. 68

4.3 **Average occupation of sites** with one (left figure) and two particles (right figure) for the exact (solid lines), Ansatz (dotted lines) and Poissonian distribution (dot-dashed lines), as a function of the filling factor ν. The system size is $M = 20$. 69

4.4 **Quasi-momentum distribution** of particles n_k (orange, left axis) and of on-site pairs $n_k^{(2)}$ (blue, right axis), with $k = 2\pi/Mn$, $n = 0, \ldots, M - 1$. Results are shown both for the exact ground state (solid lines) and the Ansatz (dashed lines). The inset shows the long-distance scaling of the correlation functions $\langle a_{\ell+\Delta}^\dagger a_\ell \rangle \sim \Delta^{-\alpha_1}$ (orange), and $\langle a_{\ell+\Delta}^\dagger a_{\ell+\Delta}^\dagger a_\ell a_\ell \rangle \sim \Delta^{-\alpha_2}$ (blue), for the exact ground state (circles, $\alpha_1 = 0.22$, $\alpha_2 = 0.83$) and the Ansatz (diamonds, $\alpha_1 = 0.24$, $\alpha_2 = 0.99$). Parameters are $M = 20$ and $\nu = 1$. 70

5.1 **Inverse effective mass** m/m^* **as a function of the dimensionless coupling constant** γ **(logarithmic scale)**. The stars ($*$) show numerical results for $N = 111$ particles, the solid curve represents the behavior in the strong coupling limit (Eq. (5.20)), and the dashed curve the behavior for a weak coupling (Eq. (5.27)). 85

6.1 **Laser-assisted, state-dependent tunneling scheme.** 97

Acknowledgements

First and foremost I would like to thank my thesis supervisor Ignacio Cirac for the guidance and support he has provided throughout the course of this work. I am especially grateful for many inspiring and instructive discussions that have widened my fundamental understanding of physics.

I am deeply thankful to Tommaso Roscilde, Marco Roncaglia, Juan José García-Ripoll, Belén Paredes, Dimitri Gangardt and Gora Shlyapnikov for their fruitful and constructive collaboration on important parts of this thesis.

I also thank Roman Schmied, Karl Gerd Vollbrecht, Miguel Angel Martín-Delgado and Stefan Dürr for their valuable advice and critical questions.

My regards extend to all my colleagues and friends at the Max-Planck Institute, especially to Eric Kessler, Heike Schwager, Valentin Murg, Birger Horstmann, Sébastien Perseguers, Leonardo Mazza, Fernando Pastawski, Matteo Rizzi, Christina Kraus, Maria Eckholt Perotti, Henning Christ and Markus Popp, for many stimulating discussions and various inspiring side activities.

Furthermore I am indebted to the Deutsche Telekom Stiftung, the Studienstiftung des deutschen Volkes, the Max-Planck Society, the European Union and the Wellness Heaven Resort & Hotel Guide for financial support.

Finally, I wish to thank Andrea and my family for the encouragement and love I have received throughout my time at the MPQ.

Die VDM Verlagsservicegesellschaft sucht für wissenschaftliche Verlage abgeschlossene und herausragende

Dissertationen, Habilitationen, Diplomarbeiten, Master Theses, Magisterarbeiten usw.

für die kostenlose Publikation als Fachbuch.

Sie verfügen über eine Arbeit, die hohen inhaltlichen und formalen Ansprüchen genügt, und haben Interesse an einer honorarvergüteten Publikation?

Dann senden Sie bitte erste Informationen über sich und Ihre Arbeit per Email an *info@vdm-vsg.de*.

Sie erhalten kurzfristig unser Feedback!

VDM Verlagsservicegesellschaft mbH
Dudweiler Landstr. 99
D - 66123 Saarbrücken

Telefon +49 681 3720 174
Fax   +49 681 3720 1749

www.vdm-vsg.de

Die VDM Verlagsservicegesellschaft mbH vertritt

Printed by Books on Demand GmbH, Norderstedt / Germany